Praise for

Spring Chicken

"Gifford skillfully navigates the many strands of aging research to create an entertaining narrative of the perils of getting old."

—*Kirkus*

"[Gifford's] survey of those who study aging and those who claim they can slow it down or stop it makes for a great read."

—*The Washington Post*

"[A] hilarious quest for the proverbial fountain of youth."

—*Outside Magazine*

"Engaging... Though he doesn't skimp on the relevant science, the tone remains accessible, even humorous, as Gifford threads his own personal journey and experiences together."

—*Publisher's Weekly*

"Gifford's wit and keen eye for interesting details will endear his work to both aging baby boomers and anyone who appreciates top-notch popular-science writing." —*Booklist*

"It's nice, then, when a book comes along and cobbles these studies together, weighing their evidence and methods against each other with appropriate skepticism. Out this week is one such book, *Spring Chicken: Stay Young Forever or Die Trying* in which author Bill Gifford investigates the rapidly advancing science of aging and the quest for an anti-aging pill." —*Daily Beast*

"Deeply researched, but also deeply entertaining, this very readable book is packed with analysis and insight, and Gifford makes an obvious truth very evident. A key to counteracting aging is, bottom line, to get up and move and live a life of purpose and value."

—*Summit Daily News*

"A thorough, thoughtful, evidence-based exploration of how and why people age differently. Sure, it's also playful and funny, but it's never flighty."

—*Winnipeg Free Press*

"As clever and humorous as its title, this deeply reported examination of what to do about getting old by veteran journalist Bill Gifford offers some prescriptive takeaway."

—*Departures Magazine*

"[Gifford] has done an excellent job of explaining the pertinent scientific literature and putting it into perspective. What is really refreshing and what distinguishes his book from other books on the subject is that he stresses the uncertainty of our present knowledge and doesn't try to tell readers what they *should* do to reach a ripe, healthy old age. I highly recommend the book."

—sciencebasedmedicine.org

"You need this book. I grabbed it like a life preserver, and that's exactly what it is. SPRING CHICKEN demolishes the worst hoaxes in anti-aging treatments—like crushed dog testicles, human growth hormone, and Suzanne Somers—and leaves you with the good news: by adopting a few easy-to-understand, easy-to-follow discoveries, you might just deactivate the time bombs in your fat cells and learn to follow in the springy, 'successfully aging' footsteps of a 92-year-old pole vaulter."

—Christopher McDougall, *New York Times* bestselling author of *Born to Run* and *Natural Born Heroes*

"SPRING CHICKEN is an utterly marvelous book—a guided tour of a fantastic, counterintuitive landscape (that happens to be your body), and also a whip-smart guide to living a longer and healthier life. With this book, Bill Gifford joins the ranks of Mary Roach and Bill Bryson as a science writer supreme, illuminating our world in a page-turning style that is as entertaining as it is enlightening."

—Daniel Coyle, *New York Times* bestselling author of *The Talent Code*

"Bill Gifford's terrific SPRING CHICKEN gives us a riveting account of the most important change of the last century—the doubling of our lifespans—and an intimate vision of what it will take to not only keep that trend going, but keep ourselves healthy and vibrant as we age."

—Steven Johnson, *New York Times* bestselling author of *How We Got to Now* and *Where Good Ideas Come From*

SPRING CHICKEN

STAY YOUNG FOREVER (OR DIE TRYING)

BILL GIFFORD

GRAND CENTRAL PUBLISHING

NEW YORK BOSTON

Grand Central Publishing
Hachette Book Group
1290 Avenue of the Americas
New York, NY 10104

hachettebookgroup.com
twitter.com/grandcentralpub

Originally published in hardcover by Hachette Book Group.

First trade paperback edition: February 2016

Grand Central Publishing is a division of Hachette Book Group, Inc.

ISBN: 978-1-4555-2743-4

For my parents

I think the most unfair thing about life is the way it ends. I mean, life is tough. It takes up a lot of your time. What do you get at the end of it? A death! What's that, a bonus? I think the life cycle is all backwards. You should die first, get it out of the way. Then you live in an old age home. You get kicked out when you're too young, you get a gold watch, you go to work. You work for forty years until you're young enough to enjoy your retirement! You go to college, you do drugs, alcohol, you party, you have sex, you get ready for high school. You go to grade school, you become a kid, you play, you have no responsibilities, you become a little baby, you go back into the womb, you spend your last nine months floating—and you finish off as a gleam in somebody's eye.

—Sean Morey

Contents

Prologue

THE ELIXIR

You are never too old to become younger.

—Mae West

In his final moments of consciousness, as the young scientist crumpled to the laboratory floor, he may have realized that perhaps covering himself with varnish was not the best idea he had ever had, experiment-wise. But he was a man of science, and curiosity could be a cruel mistress.

He had been wondering for a while about the function of human skin, so durable and yet so delicate, so sensitive to burns from sun and flame, and so easily sliced open by knives much less sharp than his surgeon's blades. What would happen, he wondered, if you covered it all up?

So, on what was otherwise a slow day at the lab, at the Medical College of Virginia in genteel Richmond, Virginia, in the spring of 1853, Professor Charles Edouard Brown-Séquard—native of Mauritius, citizen of Britain, late of Paris (via Harvard)—stripped off all his clothes and went to work on himself with a paintbrush and a pail of top-quality flypaper varnish. It didn't take long before he had coated every square inch of his naked body with the sticky liquid.

This was still an era when a scientist's primary guinea pig was generally himself. In one experiment, the thirty-six-year-old Brown-Séquard had lowered a sponge into his own stomach to sample the digestive juices within, which caused him to suffer gastric reflux for the rest of his life. Such practices distinguished him as "by far the most picturesque member of our faculty," as one of his students would later recall.

The varnish episode would only add to his legend. By the time a random student happened to stumble across him, the professor was huddled in a corner of his lab, trembling and apparently near death. His body was so brown that it took a moment for the student to realize that he was not a wayward slave. Thinking quickly, the young man frantically began to scrape off the brown gunk, only to receive a tongue-lashing from the victim, who was furious that "some obtrusive individual [had] extracted him from the corner into which the varnish had tumbled him, and, just as he was fetching his last gasp, maliciously sandpapered him off."

Thanks to that quick-witted medical student, though, Brown-Séquard would go on to become one of the greatest scientists of the nineteenth century. Today he is remembered as a father of endocrinology, the study of glands and their hormones. As if that were not enough, he made major contributions to our understanding of the spinal cord; a particular type of paralysis is still called Brown-Séquard syndrome. Yet he was far from an ivory-tower academic. He once spent months battling a deadly cholera epidemic on his native Mauritius, a lonely archipelago in the middle of the Indian Ocean. True to form, he intentionally infected himself with the disease by swallowing the vomit of patients, in order to test a new treatment on himself. (That nearly killed him, too.)

His Richmond professorship did not last the year; the Frenchman's eccentric ways and darkish skin proved too much for the Southern capital, so he moved back to Paris, to spend the remainder of his career shuttling between France and the United States. All told, he spent six years of his life at sea, which would have made his late sea-captain father proud. Yet despite his near-constant state of motion, he could not outrace old age. By his sixties, Brown-Séquard had fetched up once more in Paris, as a professor at the Collège de France. His friends included Louis Pasteur, as in *pasteurization*, and Louis Agassiz, one of the forefathers of American medicine. The poor orphan from far Mauritius was inducted into the French Legion of Honor in 1880, followed by a slew of other prestigious prizes, culminating with his election as president of the Société de Biologie in 1887, confirming his status as one of the leading men in French science.

He was seventy years old by then, and he was tired. Over the previous decade, he had noticed certain changes overtaking his body, none of them good. He had always buzzed with frenetic energy, bounding up and down stairs, talking a mile a minute, then interrupting himself to scribble down his latest brilliant idea on the nearest scrap of paper, which would vanish into a pocket. He slept just four or five hours a night, often beginning his workday at his writing desk at three in the morning. It has been suggested, by his biographer Michael Aminoff, that he may have been bipolar.

But now his once-boundless vigor seemed to have abandoned him. He had evidence, too, because he had long kept track of his body, measuring things like the strength of his muscles and keeping careful records. In his forties, he had been able to lift a 110-pound weight with one arm. Now the best he could do was eighty-three pounds. He got tired quickly, yet he slept poorly if at all, and he was tormented by constipation. So naturally, being the scientist he was, he decided to try to fix the problem.

On June 1, 1889, Professor Brown-Séquard stood before the Société de Biologie and delivered a keynote address that would forever change his career, his reputation, and popular attitudes toward aging. In the talk, he reported on a stunning experiment that he had performed: He had injected himself with a liquid made from the mashed-up testicles of young dogs and guinea pigs, which he had augmented with testicular blood and semen.

His idea was simply that something in younger animals—specifically, in their genitals—seemed to give them their youthful vigor. Whatever that was, he wanted some. After a three-week cycle of injections, he reported a dramatic turnaround: "To the great astonishment of my principal assistants," he claimed, "I was able to make experiments for several hours while standing up, feeling no need whatever to sit down."

There were other benefits. His strength seemed to have returned, as his tests confirmed: Now he could hoist a hundred-pound weight, a significant improvement, and he was once again able to write late into the evenings without fatigue. He even went so far as to measure his "jet of urine," and found that it now traveled 25 percent farther than it had prior to the injections. With regard to his constipation issues, he noted proudly that "the power I long ago possessed had returned."

His colleagues in the audience were torn between horror and embarrassment. Extract of... *dog testicles*? Had he gone mad in his old age? Later, one of his colleagues sniped that Brown-Séquard's outlandish experiment had proved only "the necessity of retiring professors who have attained their threescore and ten."

Undeterred, he made his magic mixture (now fashioned from the testes of bulls) available for free to other doctors and scientists, in the hope they could repeat his results, which some did. The reviews from his peers were still scathing. Harrumphed one Manhattan MD in the pages of the *Boston Globe*, "It is a return to the medical systems of the middle ages."

Outside the halls of academe, though, Brown-Séquard became an instant hero. Almost overnight, mail-order entrepreneurs began selling "Séquard's Elixir Of Life": twenty-five injections for $2.50, using the good doctor's name but with no other connection to him. The newspapers, predictably, had a field day; at last, they could print the phrase *testicular liquid*. A professional baseball player, Jim "Pud" Galvin of Pittsburgh, openly used the elixir in the hope that it would help him pitch better against Boston—the first recorded modern use of a performance-enhancing substance by an athlete. The old professor was even feted in a popular song:

The latest sensation's the Séquard Elixir
That's making young kids of the withered and gray
There'll be no more pills or big doctor bills,
Or planting of people in churchyard clay.

Sadly, this last line proved to be wishful thinking: On April 2, 1894, five years after his address to the Société de Biologie, Charles Edouard Brown-Séquard was dead, six days shy of his seventy-seventh birthday. Despite his fame, he had not profited one franc from his elixir. And while his fellow scientists ultimately concluded that the miraculous revival that Brown-Séquard had attributed to his "orchitic liquid" was due to little more than a placebo effect, he had kicked off a rejuvenation craze that seemed to cause even the most rational men and women to lose their minds.

The next fad was something called the Steinach operation, which promised to restore a man's vitality but really amounted to nothing more than an ordinary vasectomy. It nevertheless became hugely popular among the male intelligentsia of Europe, including the poet William Butler Yeats, who at sixty-nine had married a twenty-seven-year-old; even Sigmund Freud, so attuned to phallic states, pronounced himself satisfied with the results.

In the United States, rejuvenation fever exploded in the 1920s, when a patent-medicine salesman named John Brinkley popularized an operation that basically involved implanting fresh goat testicles into the scrota of worn-out middle-aged men. Brown-Séquard had actually tried similar experiments on dogs back in the 1870s, but even he hadn't dared try a cross-species transplant. Brinkley had no such qualms, perhaps because he was unencumbered by an actual medical education.

He did, however, own a radio station, and he broadcast nonstop testimonials about the wonders of the operation. (His station later became famous as one of the wellsprings of country music, for he also broadcast early performances by the Carter Family and a young Elvis Presley; this earned him a name-check from Lynyrd Skynyrd.)

Over the decades, he operated on thousands of patients, making himself one of the richest men in America. Meanwhile, dozens of people died on his operating table, and hundreds more were left crippled or maimed by his clumsy surgeries. And still they kept flocking to him: the tired, the worn-out, the flagging, impotent, aging men of America, and even a few brave women, desperate for one more chance at youth.

They had no idea how lucky they were, just to be alive.

Chapter 1

BROTHERS

Old age isn't a battle; old age is a massacre.

—Philip Roth

The wave reared up, green and foaming, and slammed into my grandfather. For a too-long instant, he disappeared under the water. I watched from the shore, holding my breath. I was ten years old. Finally, he staggered to his feet on the shallow sandbar, wiped the spray from his eyes, and turned to face the next rising wall of water.

Lake Michigan has days when it thinks it's an ocean, and that day was one of them. All morning long, it had been hurling five-foot swells at the beach in front of my family's old frame cottage, which my great-grandfather had built with his own hands, cheap lumber, and sheer Anglo-Saxon will back in 1919. Bodysurfing on this beach was one of my favorite things on earth, and I prayed for wavy days. Unfortunately, on this day the waves were too big, and I had been forbidden to go in the water. So I sat on the porch, sulking.

With me on the porch was my great-uncle Emerson, who was

my grandfather's older brother and, it's fair to say, not my favorite relative just then. Stiff and somewhat humorless, he only spoke to us children to scold us for running around or making noise. He didn't swim, so he couldn't watch us on the beach, which rendered him pretty much useless to us. He never joked or played, either, the way the other uncles did. He just sat there, staring vacantly out at the lake. To my ten-year-old mind he just seemed ancient, and not in a good way like a fossil or a dinosaur.

Meanwhile, out in the water, my grandfather was frolicking in the head-high waves. His name was Leonard, and even in his sixties, the old navy man still loved the rough surf. Enviously, I watched him plunge into one foaming breaker after another, emerging to wipe the water from his eyes before turning to face the next one. I adored him.

The family had gathered to celebrate his birthday, which he had mock-grandiosely dubbed St. Leonard's Day. A homemade banner proclaiming it as such fluttered from the porch railing, to the puzzlement of beach walkers. The house was a kind of landmark because it was so much older than its neighbors. It had survived the Great Depression and countless brutal winter storms, including a big one during the 1930s that had washed out the sand dune on which it had been built. Nearly all the neighboring cottages were completely destroyed. The family drove out from Chicago and repaired it by themselves, and after that it was known as The Ark.

The adults gathered for cocktails at six. It may have been closer to five. Afterward the aunts fixed dinner in the downstairs kitchen, built to hold up the house after we lost the dune. When dinner was done, the men lit a bonfire on the beach, and we kids scorched marshmallow after marshmallow until we were sent to bed, to the sound of crashing waves. It was just another beautiful childhood day at the lake, and it would stay lodged in my memory for years before I recognized its true significance.

Though they almost seemed like they came from different generations, my grandfather Leonard was a mere seventeen months younger than his brother Emerson, a slender gap bordering on scandalous for upright Midwestern Protestants circa 1914–15, when they were born. They were nearly twins, with the same genes and upbringing, and they remained very close throughout their adult lives. Yet their fates could hardly have been more different.

Still that image haunts me: Emerson in his rocker on the porch while his only-slightly-younger brother was out there ducking major waves. Not too long after that day, Emerson began showing signs of frailty that would eventually send him to a nursing home, where he died at just seventy-four. Meanwhile, my grandfather's idea of retirement was to buy a small citrus orchard in the mountains north of San Diego, where he toiled alongside the migrant farmhands until his mid-seventies. He was still going strong when a random infection felled him in his eighty-sixth year.

The difference between the two brothers was at least partly the result of one unlikely factor: religion. Like my great-grandparents, Emerson and his wife were devout believers in Christian Science, which is a misnamed faith if ever there was, because its followers actually reject medical science in the belief that human ailments can be healed through prayer. So they almost never went to the doctor, for anything. As a result, Emerson had piled up biological damage like a Cadillac in a demolition derby. A succession of skin cancers that he refused to treat had eaten away at his left ear, leaving it deformed and cauliflower-like. Later, he suffered a series of minor strokes that also went unattended. Every time he had an infection that could have been cleared with antibiotics, but wasn't, that took a toll on him, too.

My grandfather had shed his Christian Science beliefs early, at the insistence of his wife, and his most consistent religious observance was a steadfast devotion to the daily cocktail hour: one Scotch-based beverage on the rocks at 6 p.m. sharp every day. He availed himself of modern medical care, which had made crucial advances against infectious illnesses, and even heart disease and cancer. Just as importantly, he had quit smoking in 1957 (unlike his brother he was an early adopter of healthy habits, such as eating fiber daily), and he got daily exercise in the form of vigorous and often highly ambitious gardening projects, which he worked on every day before cocktails. The result was that he enjoyed a longer life—and a much longer *healthy* life—than his brother.

Public-health experts now call this *healthspan*, one's span of healthy years, and it will be an important concept in this book: While my grandfather's *life*span was only about twelve years longer than his brother's, his *health*span was at least thirty years greater. If I've done my job, *Spring Chicken* will help you understand how to end up more like my grandfather, with his long healthy life, and less like his unfortunate brother.

Decades later, on another perfect summer day, I found myself again sitting on the porch of The Ark. It had been a long time since I'd visited. My grandfather's generation had moved on, and the house had been sold to a distant cousin. We didn't go there much anymore, so this was a rare treat, a return to the site of some of my happiest childhood memories. Only now I was in my early forties, and naturally I had been thinking gloomy thoughts about getting older.

This was in part thanks to my thoughtful work colleagues, who had marked my fortieth birthday by giving me a cake adorned with a single candle. Shaped like a tombstone, it read:

RIP

MY YOUTH

Which was awfully kind of them. But it was also rather brutally true: In the media world in which I've worked all my professional life, forty *is* considered old. Even though you aren't actually old—far from it—our culture nonetheless labels you middle-aged. Demographically undesirable. On the way out, career-wise. Possibly even an AOL user. My own mother had already pronounced me "no spring chicken."

She did have a point. Inside, I could tell something was changing. I'd been more or less athletic since college—sometimes more, sometimes less—but lately I'd noticed that it had become a lot more difficult to keep in shape. If I took a few days off from running or cycling or going to the gym, my muscles would turn to Jell-O, as though I'd been sitting on the couch for weeks. When I finally did get out for a jog, I'd feel the unmistakable bounce of nascent man-boobage.

Hangovers now seemed to last for days, my wallet and my keys liked to go AWOL, and as for reading a restaurant menu by romantic candlelight, forget about it. I seemed to be tired all the freaking time. A handful of friends had already died of cancer, or come close. In idle moments, I found myself dwelling more and more on middle-aged regrets, stuck on the idea that my best years were behind me, and that God was checking his watch. Right on schedule: Some scientists believe that the woes of midlife reflect the fact that we have reached a kind of biological "tipping point," where the damage of aging has begun to outpace the ability of our body and our mind to repair themselves.

When I went in for a physical exam, somewhere around age forty-three, I learned that I had mysteriously gained fifteen pounds, and my cholesterol levels now approximated those of chocolate milk. For the first time ever, I had the beginnings of a beer belly, which shouldn't have been surprising since I love beer, but it bummed me out nonetheless. All of this my doctor chalked up to "normal aging." She smiled as she said it, as if it were nothing to worry about, and certainly no reason to take action. Nothing to be done, her slight shrug said.

Really? I wanted to know more. Like, can we make it stop? Or at least slow down? A little? Please?

Finding a "cure" for aging, a way to defeat death, has been the dream of humankind literally since we began writing down our dreams. The oldest existing great work of literature, the nearly four-thousand-year-old *Epic of Gilgamesh*, in part chronicles a man's quest for the elixir of eternal life. He actually finds it, in the form of a mysterious thorny plant that he retrieves all the way from the bottom of the sea, only to have it stolen by a serpent (spoiler alert). "When the gods created man they allotted him death," the hero Gilgamesh is told, "but life they retained for their own keeping."

Staying young, or at least looking young, has been much on our minds. One of the oldest known medical texts is an Egyptian papyrus dating from circa 2500 BC that contains a "Recipe for

Transforming an Old Man into a Youth." Unfortunately, the recipe turns out to be a face cream made from fruit and mud, probably not all that different from the pomegranate- and melon- and milk-infused "anti-aging" creams that Americans spent an estimated eleventy bajillion dollars on last year. My favorite is a seaweed-based potion called Crème de la Mer that sells for more than $1,000 a pound; the *Daily Mail hired* a cosmetic chemist who analyzed the ingredients and determined that they actually cost more like $50.

When *Gilgamesh* was written, relatively few humans lived long enough (or well enough) to die of old age; life expectancy hovered around twenty-five years, as it had for millennia. On the day you are reading this, ten thousand Baby Boomers will celebrate their sixty-fifth birthdays. Tomorrow, another ten thousand will crank up the Jimmy Buffett and float across the Rubicon of "old age"—and so on and so on for the next two decades. At this rate, we will run out of birthday candles well before 2060, when the number of Americans older than sixty-five will have doubled to more than ninety-two million, making up 20 percent of the U.S. population. For comparison's sake, over-sixty-fives make up just 17 percent of the population of Florida right now.

The entire planet is turning into Florida. There are more older people on earth right now than ever in history, even in recently "developing" nations like China, where the one-child policy has skewed the population balance in a breathtakingly short period of time. For most of human history, the age distribution of the human race has resembled a pyramid, with a great many young people at the base, and relatively fewer oldsters as you move up toward the peak. Now, as lifespans get longer and birthrates get smaller, the industrialized countries have become top-heavy with old folk, more like mushrooms than pyramids. According to the

Nikkei newspaper, Japan will soon sell more adult diapers than diapers for children. Instead of succumbing to tuberculosis or polio or the plague, as in previous generations, these "new old" will die of heart disease, cancer, diabetes, and Alzheimer's—the four horsemen of the geriatric apocalypse.

These chronic diseases have become so common as to seem inevitable. Four out of five American sixty-five-year-olds are now on medication for one or more long-term ailments—for high cholesterol, blood pressure, diabetes, and sundry other complaints. Increasingly, our old age is a highly medicated one, which means that we are likely to spend the latter decades of our lives as patients—that is to say, as sick people. Public-health experts call this the period of morbidity, the portion of our lives when we suffer from chronic disease. Right now, for most people, that period consists of, basically, the second half of their lives, which is a scary thought. Scarier still is how much these legions of aging Baby Boomers are going to cost to keep around, with their medications and knee replacements and artificial heart valves—and how lousy many of them are still going to feel.

If there were ever a time when humanity needed the magic flower of Gilgamesh, this would be it.

As Montaigne observed, the real cruelty of aging is not that it kills an old person, but that it robs a young person of his or her youth. That is the greater loss, he wrote. The only mercy is that it works slowly, almost imperceptibly. Nevertheless, he wrote, Nature "step by step conducts us to that miserable state... so that we are insensible of the stroke when our youth dies in us, though it be really a harder death than the final dissolution of a languishing body, than the death of old age."

Though I missed the Baby Boomer cutoff (1964) by three years, I did share in their grand generational delusion, that we would somehow never get old. Aging was something that had happened to old people, our parents and grandparents. We, somehow, would be immune. So much for that, obviously, but what made aging real for me, finally, was not my parents hitting seventy, or even my own impending cage-match with middle age; what brought it home, at last, was what happened with my dogs.

There were two of them, a matched pair of redbone coonhound mixes, the Southern breed featured in the children's classic *Where the Red Fern Grows.* I'd had Theo from puppyhood, and Lizzy since she was very young, and now they both qualified as canine senior citizens. The interesting thing was that while Theo had stayed sort of puppyish, Lizzy had gone gray in the muzzle at seven or eight, and had developed a stiff-legged, lady-truck-driver walk. People would approach us on the street and ask, with no regard for her considerable vanity, "Is she the mother?"

Nope: They were brother and sister, born in the same litter. But they looked so different, it was like my grandfather and Emerson all over again: One seemed so much older, yet they were exactly the same age. Only with the dogs, there was no obvious explanation, like Christian Science. They had basically the same genes, and had eaten the same food and gone for the same walks since they were young. Like my grandfather and his brother, they could not have been more similar—or more different.

Everyone has noticed this, how people seem to age at vastly different rates. We go to a school reunion, and some classmates have turned into their parents, while others look like they just got home from Beach Week The corollary to this, of course, Is that everyone thinks they look younger than they really are. What makes the difference? Is it only "good genes," as most people seem to think? Or is it something you can control, like what you eat? How much you moisturize? Answering this big question—why some people age more slowly than others—will be a key mission of this book.

With Theo and Lizzy, I chalked it up to random chance—which actually does play a significant role in aging, scientists believe. But that wasn't quite it, and as it turned out, appearances were deceiving. One October Sunday, I came home from a bike ride to find Theo waiting on the porch of our cabin in Pennsylvania, all excited. He used to love racing with me on the trails, and even now that he was nearly twelve, he was still up for a quick trot around the block. So I opened the gate, and he cantered alongside me for a lap, then two, then three. He seemed fine, ready for more, so it was a shock when I took him to the vet four days later and found out that he had cancer.

Our vet is a kindly man named Tracy Sane, a country boy marooned in Manhattan, and whenever he saw the two redbones, he would get a little wistful and say something like, "Those are *real* dogs." I'd brought Theo in to have a small skin growth removed, which should have been no big deal. The surgery would require him to go under anesthesia, so Dr. Sane donned his stethoscope to listen to his heart. As he worked his way down Theo's chest, his expression darkened. "Theo's got a bit of a heart murmur," he said.

The murmur meant that Theo's heart was enlarged, and weakened. It happens to humans as well, and is one of the most common signs of aging. And it usually means there is something else wrong. The chest X-ray revealed what it was: The space where his spleen and liver should have been was occupied by a large, fuzzy blob, about the size of a toy Nerf football. "This," said Dr. Sane, "is a problem." He called it a "splenetic mass," which was a soft way of saying "tumor." It needed to come out—if it could be removed safely, he said. We made an appointment to come in first thing Monday morning. "Theo's looking at a tough road," he warned grimly.

Over the weekend, my girlfriend Elizabeth and I tried not to think about Theo and his tumor. The news was all about a hurricane called Sandy that was preparing to slam into the city. It was supposed to be one of the strongest storms ever to hit New York. On Saturday we walked to the neighborhood farmer's market, where Theo and Lizzy tugged us toward their favorite stand, the one that sold turkey sausages and gave free samples to dogs. Then we snuggled up on the couch with the TV on, watching the tall sailing ship *Bounty* as she sank off North Carolina. Sandy was coming.

On Sunday we hunkered down for the storm, reading the paper and drinking coffee and then switching to wine. After dinner, we tried to get the dogs out for one last walk, but Theo wouldn't go. This wasn't unusual. He hated storms, and he had been known to hold out for hours rather than venture out in the rain to pee. He was a stubborn guy, and there was no dragging him. I gave him a sort of doggy massage to try to relax him, rubbing up and down his back as he lay in his bed. But we didn't think there was anything terribly wrong, other than the weather. The next morning, when the storm had passed, we'd take him in for his operation. He was three weeks short of his twelfth birthday.

But Theo had other plans, and they didn't include surgery. We found him before dawn, lying beside his bed, still warm except for his lips. I closed his eyes, Elizabeth pulled a blanket around his body, and we wept together.

* * *

In the weeks after Theo died, more than one friend confided that they had cried harder over the death of a dog than when their own fathers had passed away. It's not that they loved their fathers any less (or at least, not entirely that). But our parents grow old in slow motion, and we expect it. There's something about a beloved animal's short life and quick passing that hits too close to home. It reminds us too much of our own tenuous lease on this existence. In Theo's lifetime, I had gone from being a still-pretty-young man, with thirty just in the rearview, to one who was no longer quite so young, even pushing fifty.

I was so old that I was actually working on a book about aging. Theo's death pushed me into overdrive. Now I wanted to know *everything* about aging, this universal but still little-understood process that affects practically every living thing. I decided to approach it as a reportorial investigation, following the evidence wherever it led. I would read every study, every book on the subject of aging that I could find. I'd worm my way into the underfunded laboratories where the hard science was done, and I would ferret out the leaders of the field. But I would also seek out the mavericks, the rebels of science, the ones who had the courage to push novel insights, regardless of current dogma or fashion. I'd also look for the older people who are showing the way to the rest of us: the ones who are pole-vaulting in their seventies, thought leaders in their eighties; even picking stock-market winners past a hundred.

I had big questions: How does time transform us? What was happening to me, as I slid into middle age, and beyond? How was my mid-forties self different from my teenage self? What would change between forty and seventy? For that matter, why is my ten-year-old niece "young," but my twelve-year-old dogs are old?

What is this invisible force called aging that affects everyone I know? Everyone reading this? Everyone who has ever lived?

More to the point: How much of aging is under our control, and how much determined by fate, or random chance? My motivation was personal. Straight up, I wanted to hang on to my youth, or what was left of it, for as long as possible. I want to end up like my grandfather, diving into the waves and pruning fruit trees in his old age—and not bound to the rocking chair, like his poor brother Emerson.

And while I'd feared, early in my research, that I'd only learn a bunch of depressing stuff, that turned out not to be the case at all. Scientists are discovering that aging is far more malleable than we had ever thought—that, in effect, it can be hacked. You don't have to endure your grandfather's old age (or in my case, my great-uncle's). How well you grow old is at least partially under your control. Two of the major diseases of aging—cardiovascular disease and diabetes—are largely avoidable, and even reversible in some cases. A third, the dreaded Alzheimer's disease, may be up to 50 percent preventable.

The story of the dogs told me that there is more to longevity than simply whether or not you go to the doctor and get a weekly facial. The mystery goes much deeper than that. What's really cool and surprising, though, is how many aspects of aging can be modified, even delayed, at the cellular level. Science has discovered secret longevity-promoting pathways and mechanisms, embedded deep within our cells, that can help beat back or slow down some of the effects of aging—if we can only figure out how to unlock them. Some of these evolutionary pathways are so ancient that we share them with the lowest life-forms, such as microscopic worms and even yeast; others we are only beginning to identify, through the enormous power of genomic sequencing.

Already, we know that certain genes seem to be linked to extreme longevity and good health, and hundreds more such genes are on the brink of being discovered. Some of them may even be able to be triggered, or mimicked, by drug compounds that are already in the research pipeline. But not everything is pie-in-the-sky: Major longevity-promoting mechanisms, hard-wired into our biology, can be triggered right now, by simply going out for a short jog, or even just by skipping a meal or two. A little bit of knowledge and prevention, it turns out, may even make the difference between bodysurfing your way through the rest of your life, and spending it on the rocker on the porch.

Chapter 2

THE AGE OF AGING

The days of our years are threescore years and ten; and if by reason of strength they be fourscore years, yet is their strength labor and sorrow; for it is soon cut off, and we fly away.

—Psalm 90:10

The divergent fates of my great-uncle Emerson and my grandfather Leonard reflect the vast increase in human lifespan that has occurred over the last century. Emerson lived his life like a man of the late nineteenth century, when Christian Science was founded: a brief, flourishing youth, followed by a long, painful decline, beginning in middle age. Frankly, it's sort of amazing that he made it past seventy. My grandfather, meanwhile, was very much a twentieth-century man: Forward thinking and science-minded, he availed himself of the best that modern medicine had to offer. No surprise that he lived a much longer and healthier life than his brother.

And yet both men, even Emerson, had vastly outlived their predicted life expectancies at birth. When they were born, in 1914–15, a typical white American male could look forward to about fifty-two years on this earth. The leading killer of Americans, then as now, was heart disease—which had only just displaced tuberculosis and pneumonia, in retreat thanks to the advent of antibiotics. Influenza would briefly top the charts during the pandemic of 1918, but for the first time in history, more people were dying from a disease of aging than from any other cause. The Age of Aging had begun.

Today American males enjoy a life expectancy of about seventy-seven years, with another five bonus years for women, according to the World Health Organization. Globally, though, that's nothing to brag about: We're only ranked thirty-second, behind Costa Rica, Portugal, and Lebanon, despite spending far more per person on health care. And we keep falling behind: For some subgroups of the American population, life expectancies may already have begun to decline. Meanwhile, according to some estimates, half of all German children born this year will live to see their 105th birthdays.

This explosion in longevity has no precedent in human history. Take a walk around an old cemetery sometime, and read the headstone dates: You'll find a tragic overrepresentation of infants and children, and young women who died in childbirth, while the luckier men generally lived to see their forties, and a few exceptional individuals made it past seventy—their biblical allotment of threescore and ten. It was still possible to live a very long time: The first English child born in Massachusetts in 1621, a girl named Elisabeth Alden Pabodie whose parents came over on the *Mayflower*,* managed to live for nearly a century, dying in 1717 at the age of ninety-six. Back then, particularly in the rugged realm of the Massachusetts Bay Colony, growing old was

* And who happens to be—huzzah!—an ancestor of mine.

an accomplishment, not an affliction. As Montaigne put it, "To die of old age is a death rare, extraordinary, and singular, and, therefore, so much less natural than the others; tis the last and extremest sort of dying."

Things began to change in the mid-1800s, with the appearance of urban sewers and semi-modern medicine; just the widespread adoption of hand washing by doctors reduced death rates enormously. In 1881, for example, President James Garfield died not from his assassin's bullet, but from the massive infection that his dirty-fingered doctors gave him. Deaths in childbirth, once commonplace, became more and more rare thanks to the miracles of anesthesia, antibiotics, and the cesarean section, without which both I and my mother would surely have died from the trauma of bringing a nine-pound, eight-ounce child into the world. As clean water became more available (and raw sewage more distant), as medicine made progress against infectious disease, and as infant mortality plunged, life expectancies climbed rapidly. And more people than ever experienced the bizarre, inexplicable natural phenomenon called aging.

If you happen to stroll through Westminster Abbey in London, you might spot a rather extraordinary marble gravestone in the south transept. It marks the resting spot of one Thomas Parr, who according to the marker had lived for 152 years and through the reigns of ten kings. That is not a typo. Parr was a laborer in the Shropshire countryside who was famed for having lived well more than a century; indeed, he had reputedly fathered a child at the age of 122. Some earl heard about Parr and invited him to the court of King Charles I in 1635, where he enjoyed a brief celebrity that included having his portrait painted by Peter Paul Rubens.

But his ride on the celebrity gravy train was cut short when, after a few weeks' exposure to disease-ridden London and its horrific pollution, he died.

Old Parr would have been the oldest human ever to have lived—if his claimed age were even remotely close to the truth. Doubts began to arise not long after his autopsy by famed surgeon William Harvey, who noted that his internal organs were in rather good condition for being a century and a half old. Notwithstanding the age on his tombstone, modern scholars now believe "Old Parr" was actually the grandson of the *original* Old Parr, and the title was simply passed on down the line. Birth records in Shropshire were pretty spotty in the 1500s, so who can know for sure?

More recently, in the 1960s, it was claimed that residents of the Abkhazia region of the then Soviet Union, deep in the Caucasus Mountains, also routinely lived past the age of 140. Their longevity was attributed to their consumption of yogurt, which has been hugely popular ever since, despite the fact that their claims have been thoroughly debunked. Just in the last few years there's been a resurgence of highly wizened individuals popping up in places like Bolivia and rural China, claiming to be 125 years old or more. One thing these wrinkled shysters all have in common with Old Parr is that they too lack trustworthy birth certificates, so their claims cannot be verified.

The longest-lived documented human being in history was an otherwise unremarkable Frenchwoman named Madame Jeanne Calment, who was born in Arles in 1875 and claimed to have met Vincent van Gogh in the art shop of her uncle. (Not a nice man, in her opinion.) When she was around eighty, Madame Calment made a deal to sell her apartment to a lawyer friend who was then in his late forties. Under the deal, common in France, the buyer would pay her 2,500 francs per month for the rest of her life, and would take possession of the place after she died. There was only one hitch, which was that she failed to die, year after year after year. She rode a bicycle until she was 100, and smoked until she was 117; quitting might have been a mistake, because she only lasted 5 more years before she gave up the ghost at the age of 122. By that time, the poor schnook himself had died, having paid her twice what her place was actually worth.

"I've only ever had one wrinkle," she famously said, "and I'm sitting on it."

So that's life*span*. Nobody has topped Madame Calment, before or since. Period.

Life *expectancy*, on the other hand, is a statistical prediction of how long a baby born this year is likely to live, based on a dull-seeming document known as the life table. To you and me, the life table looks like an eye-glazing compendium of random numbers, about as exciting as the phone book. It lists current mortality rates—that is, the risk of dying for individuals at every single age, over the last year. So, for example, the chance that a forty-year-old American woman would die in 2010 was 1.3 in 1,000, or 0.13 percent; for a sixty-year-old female, it was 6.5 out of 1,000, a five-fold increase. If you take our hypothetical infant, and march her through this statistical gauntlet, you'll come up with her average life expectancy.

Demographers treat the life table with Talmudic reverence. It is the foundation on which the insurance industry and the retirement system are built. And it also affords a sort of window into the future. According to the life table compiled by the U.S. Social Security Administration, and used as the basis of its online calculator, a forty-seven-year-old white American male (that is, me) can expect to live for another thirty-five years. That gets me to eighty-two. Not bad, but not even as good as my grandfather. So I sought a second opinion. I downloaded an app (for real) called Days of Life that also purports to calculate one's remaining life expectancy, based on gender, age, and country of residence. Unfortunately, it said I had even *less* time left, more like thirty years—and for the next several weeks, my phone buzzed with daily reminders: "You have 10,832 days of life remaining..."

Needless to say, I deleted it. In the real world, we can't know whether we'll die at eighty-two, sixty-two, or 2 p.m. tomorrow afternoon. And, luckily, the only reliable thing about life-expectancy predictions, like weather forecasts, is that they will change.

Back in the 1920s, a prominent American demographer named Louis I. Dublin, who was chief actuary of the Metropolitan Life Insurance Company, declared that average human life expectancy would peak at precisely 64.75 years—coincidentally, just three months shy of the official retirement age of 65, designated in the Social Security Act of 1933. Back then, a typical sixty-five-year-old must have looked, felt, and smelled rather old. But it was by no means the limit. When Dublin was informed that women in New Zealand were already topping sixty-six years, he revised his estimate upward, to nearly seventy. But that also proved to be too low; even my poor great-uncle Emerson beat it.

Around the world, life expectancies have been going up relentlessly for nearly two centuries. About a decade ago, another noted American demographer named James Vaupel compiled all of the known—and reliable—historical lifespan statistics he could find, going all the way back to eighteenth-century Sweden, which kept excellent birth and death records. For each year, Vaupel and his coauthor Jim Oeppen, and their Herculean research team, identified the country where people were living the longest, according to available data—the lifespan leaders, if you will. To their astonishment, it plotted out to a straight, unbroken line, ascending steadily as an airliner out of JFK.

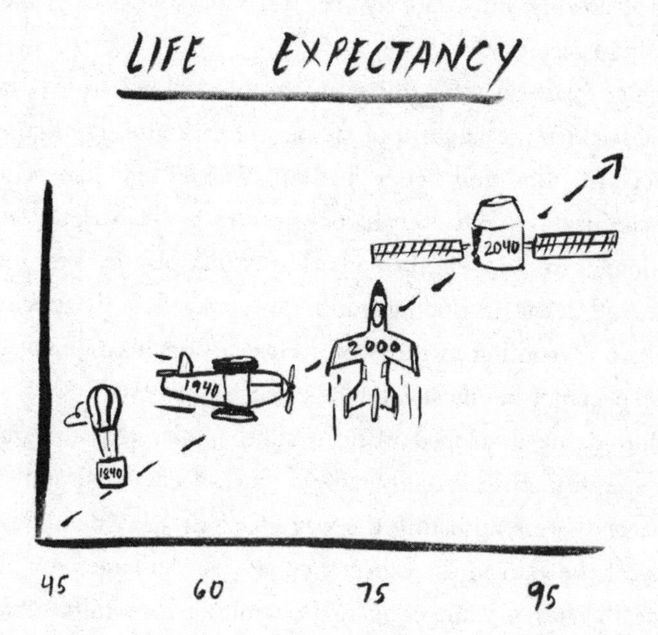

Beginning in about 1840, Vaupel's graph showed, average female life expectancy in the world's longest-lived country has increased at a steady rate of about 2.4 years per decade. And while the status of leading country has changed hands a few times, from Sweden to Norway to New Zealand to Iceland to now Japan, one thing has held true: Every four years, humans have steadily gained one extra year of potential life expectancy. Or if you prefer, every day buys us another six hours.

"The straight line absolutely astonished me," says Vaupel, from his office at the Max Planck Institute for Demographic Research in Germany. "The fact that it's held for two centuries is really amazing." Not only that, but the line plowed through many smart people's predictions that lifespan would plateau, from Dublin's to the various UN agencies to those of rival demographers, with no sign of slowing. Provocatively, he titled his study "Broken Limits to Life Expectancy."

The explanations for this relentless rise in lifespan invariably come back to the handful of factors we have already discussed: better sanitation and better medical care. Things like penicillin, sterilization, and even blood-pressure medicine have let us live longer by escaping the early deaths that plagued our ancestors. And in the developing world, this change is still happening: Globally, according to the World Health Organization, average life expectancy has increased by six years since 1990.

But in the developed world, Vaupel argues that the steady increase in lifespan actually reflects much deeper environmental changes that are affecting the way all of us age. "Before 1950, most of the gain in life expectancy was due to large reductions in death rates at younger ages," he wrote in his seminal *Science* paper in 2002. "In the second half of the 20th century, improvements in survival after age 65 propelled the rise in the length of people's lives."

It started with better medical technology: The mere fact that former vice president Dick Cheney is still alive, after his multiple heart attacks and surgeries, has to count as a marvel. Even if we're not getting new heart valves put in, we all enjoy cleaner water, cleaner air, better housing, and fewer mass epidemics than even fifty years ago. That helps explain why my grandfather's brother Emerson made it into his early seventies without any medical care at all: His world was much cleaner and safer than that of his ancestors. In fact, if he hadn't smoked—his one break with Christian Science orthodoxy—Emerson might have survived almost as long as his brother.

Indeed, widespread smoking bans have reduced everyone's exposure to tobacco smoke, a potent carcinogen, likely pushing up life expectancy even farther (although a few puffs didn't seem to hurt Madame Calment). Thanks to our ever-more-protected environment, Vaupel argues, we not only escape early death, but we are actually aging more slowly than our dirty, uncomfortable, smoke-breathing, disease-battling ancestors. "Lifespan is amazingly plastic," he says. "Seventy-year-old people today are about as healthy as sixty-year-olds were a few decades ago. They get disease and disability later; those bad five years at the end of life are now occurring at age eighty or eighty-five instead of age seventy."

Old isn't so old anymore, in other words. People are living less like Emerson, who was already ancient by the time he turned sixty, and more like my grandfather, who stayed relatively youngish into his seventies. The boundary of "old" keeps getting pushed out, by people like Diana Nyad, who swam from Cuba to Key West at the age of sixty-four, just months shy of traditional retirement age. "Sixty is the new forty" was practically her mantra. She's not that much of an outlier, either: Two of my regular bike-riding partners are guys who are on Medicare, and they make me suffer to keep up with them. And yet when Humphrey Bogart played the world-weary, decrepit-seeming Rick in *Casablanca*, filmed in 1942, he was all of forty-two years old. (Maybe it was all the smoking?)

If sixty is the new forty, then ninety-five might also be the new eighty: A recent Danish study of cognitive aging showed that the current crop of ninety-five-year-olds had reached that age in markedly better possession of their marbles than the cohort just a decade older than them. Vaupel, and others, believe that these older people are actually aging more slowly than people in previous generations. "What has become apparent over the last thirty years is that a completely new and completely unforeseen driver of the continuing increase in longevity has emerged," says Thomas Kirkwood, a prominent biologist at the University of Newcastle in England who heads up a study of the "oldest old," those older than eighty-five, "and that is the fact that people are reaching old age in better shape than they ever did before."

But can this two-century increase in lifespan continue? Will the Vaupel slope keep climbing?

Not everyone thinks so, and one leading expert believes human longevity is about to take a huge step in the wrong direction.

Jay Olshansky met me at the door of his house in suburban Chicago, and we drove to a popular hot-dog joint called Superdawg, because if there's one thing Chicago does well, it's cased-meat products. Olshansky admitted to loving hot dogs, and although he claimed he rarely ate them, he seemed to know a great deal about where to get the best ones. "As long as you don't eat like this every day, you're fine," he assured me as we pulled into the parking lot.

Which was interesting, because one thing he is known for is his firm conviction that things like "healthy lifestyles" don't ultimately affect longevity all that much. In the life-expectancy business, he is known as a strong skeptic. On the drive over, he was complaining about a billboard from the Prudential financial-services company that warned, "The first person to live to be 150 is alive today. Better be prepared."

"They're using made-up numbers," he fumed. "It has no basis in science."

Worse, if Prudential is right, then Olshansky will have lost a major bet. In 2000, he made a wager with a colleague, an evolutionary biologist named Steven Austad (whom we'll meet later on). Austad bet that in the year 2150, there will be at least one 150-year-old living on earth—in other words, that Prudential is right. Olshansky said no way. They each put up a symbolic $150—but thanks to his shrewd investments in gold, Olshansky bragged, their original $300 has grown to more than $1,200. By 2150, if this rate of return continues, the pot will be worth around $1 billion, which he fully expects his great-grandchildren to collect.

Olshansky believes that Madame Calment's 122 years represents the *upper* limit of human lifespan, a limit programmed into our genome and perhaps our very biochemistry. And that maximum figure has not changed; if anything, the old Frenchwoman was a bit of an outlier. Nobody has come close since she died in 1997. As of October 2015, the world's oldest person is 116-year-old Susannah Mushatt Jones, an African-American who lives in Brooklyn and eats bacon, eggs, and grits every day.

As for *average* life expectancy, Olshansky expects that to plateau around eighty-five, for most of the world—and, if anything, to begin to decline in some countries, such as ours.

But what about the Vaupel chart, the one that shows life expectancy climbing skyward?

"It's a fantasy, a pure fantasy," he growled, between bites of meaty goodness. He explained his reasoning: "If we extrapolate historical records for running the mile, using the same methodology, you'd come to the conclusion that in a couple hundred years from now, people will be running a mile instantaneously. Which is ridiculous."

Of course it is, although there is one important difference: Record times in the mile are getting shorter, while lifespans are getting longer. "There's a reason why you can't run a mile in zero, but there's no limit to how long you can live," Vaupel insists. And nobody is arguing that lifespans will someday be infinite. (Well, actually, one guy *is* arguing that, and we'll meet him soon.)

The Olshansky-Vaupel debate has gotten so heated and so personal that for a while, the two men would take pains to avoid attending the same conferences, lest they inadvertently run into each other. But at the core of their rivalry lies an important question: How flexible, exactly, is human longevity? What are the limits, if any?

Olshansky's basic point is worth investigating: "There are biological forces that influence how fast we can run, and biological forces that limit how long we can live," he insisted. "It's like putting air in a tire," he continued, breaking out another user-friendly analogy. "When you start pumping, it's easy, but as the tire fills up, it gets harder and harder."

For example, he said, even if we somehow cured half of all fatal cancers—the second-leading cause of death in the United States—average life expectancy would rise by a little more than three years. That's all. And even if we managed to cure heart disease, cancer, *and* stroke, the top three killers, we'd still only earn about ten years—a substantial jump, but one that still puts us short of the century mark. "You don't come close to 100," he said, "and 120 is even crazier, by several orders of magnitude."

But plenty of his colleagues would disagree—starting with Vaupel, who gleefully notes that his famous slope has already blown past Olshansky's predicted life expectancy limits. In 1990, Olshansky had confidently declared that life expectancies would soon top out around age eighty-five. Within a decade, though, Japanese women were already pushing eighty-eight. Men and women in Monaco, the world's wealthiest "nation," are already butting up against the threshold of ninety.

"You can think of the slope as the frontier of possibility, the frontier of what humans can do in terms of achieving life expectancy," Vaupel had told me.

Yes, Olshansky replied, the frontier is one thing; but how people are actually living, and more important dying, is quite another. If anything, he believes lifespans will soon begin to *decline* in many areas of the developed world—something else that has rarely been seen in modern history, except for times of war and widespread disease. "There are lots of things you can do to shorten your life, but lengthening it is a different issue," he said.

One good way to shorten your life, statistically, is by becoming obese. Olshansky believes that the epidemic of tubbiness that began in the United States in the early 1980s has already slowed the growth in life expectancy. A third of the population is officially obese, with another third clocking in as overweight, meaning with a body-mass index (BMI) between twenty-five and thirty. As a result, in nearly half of all U.S. counties, many of them in the rural Southeast, female mortality rates have already started rising again, after dropping for decades. In some parts of Mississippi and West Virginia, life expectancy for men and women is lower than in Guatemala.

The problem is not limited to country folk: A recent *Journal of the American Medical Association* (*JAMA*) study showed that the Baby Boom generation is the first in centuries that has actually turned out to be less healthy than their parents, thanks largely to diabetes, poor diet, and general physical laziness. The percentage of women who say they never engage in physical activity has tripled since 1994, from 19 percent to nearly 60 percent. Younger generations are faring even worse, succumbing to obesity at ever younger ages, particularly women between nineteen and thirty-nine. Another study, compiling data from autopsies of people who died in accidents prior to age sixty-four, showed that their cardiovascular risk factors were actually much worse than expected, meaning the long-running improvement in Americans' heart health that has been going on since the 1960s appears to have stalled. For these folks, sixty is not the new forty; forty is the new sixty.

"The overall health of the population is growing worse, not better," Olshansky asserted. "And it's getting worse faster than we thought." By his estimate, overall U.S. life expectancy could decline by between two and five years over the next couple of decades—a steep drop away from the Vaupel slope.

It's not just in the United States, either: Obesity and diabetes rates are soaring in places like India, and even in the Japanese island of Okinawa, famed as a "Blue Zone" because of its large numbers of centenarians. Thanks in part to the heavy U.S. military presence, middle-aged Okinawans eat a fast-food-heavy diet that is turning them into some of the least healthy people in Japan. The Blue Zone is turning into a red zone.

Everyone ages, but not everyone ages equally. In poorer countries, poorer states, and poorer neighborhoods, life expectancies tend to be far shorter than average; one study of London residents even showed that where you get off the Tube can make a huge difference in how long you live. Lower education levels, Olshansky says, are an even stronger predictor of early mortality. Still other research suggests that the educational level of one's mother is a key determinant of late-life health. "America is diverging," he said. "We're going to see breakthroughs in longevity for some, along with a drop in life expectancy for large subgroups of the population."

He eyed the last bite of my WhoopskiDawg, a massive Polish sausage buried in mustard and grilled onions. "How was that?"

"Jay Olshansky is a smart guy, and a friend of mine," says Aubrey de Grey, his extravagant beard twitching with each sharp syllable. "But he says some *incredibly* stupid things. I mean, it's almost embarrassing."

The verdict is rendered more damning by de Grey's clipped, British-boarding-school accent, a voice he has used to dismiss, dispute, and intimidate his critics and debate opponents for more than a decade and a half. We'd been chatting on the scruffy sofa in the offices of his foundation when an emergency struck: He ran out of beer. So we decamped to a nearby pub, which is relatively empty at four in the afternoon here in Mountain View, in the healthy, industrious heart of Silicon Valley.

He was referring to Olshansky's insistence that lifespan itself is finite, somehow programmed into our genome with the immutability of a Biblical Commandment: *Thou Shalt Not Live Longer than 120 Years.* For Aubrey, potential human lifespan doesn't end at 120; rather, that's just a beginning. He is famous for, in reverse order, his beer intake (prodigious, yet somehow not debilitating); his beard (Duck Dynasty meets Osama bin Laden); and his views on aging, which were once considered extreme but are increasingly, if grudgingly, accepted by some mainstream scientists.

With his London Fog complexion, red-rimmed eyes, and heroin-addict build, the fifty-two-year-old de Grey looks distinctly out of place in the robust California sunshine, like a religious hermit on a cruise ship. In fact, he's anything but reclusive: He's just returned from a TED speakers' reunion, before flying back to England. He maintains a grueling schedule of meetings, lectures, conferences, and interviews like this, which he conducts while also answering emails at the rate of about one every five minutes.

You might have seen him on *60 Minutes* a few years ago when, pint in hand, he informed Morley Safer that some people who were alive now would live to be a thousand years old. In an article in a scientific journal published around the same time, he went even farther, claiming that people born at the end of this century might be able to enjoy lifespans of five thousand years or more. That is roughly equivalent to someone from the Bronze Age living long enough to open a Facebook account.

This kind of talk drives Olshansky completely bonkers—"He makes up numbers depending on who he's talking to!" he sputtered—but de Grey answers such complaints with a simple argument: "Just because something hasn't happened yet, doesn't mean it won't *ever* happen."

Example A: powered human flight, first proposed by Leonardo da Vinci circa 1500, realized by the Wright brothers some four centuries later, propelled by jet engines just fifty years after that, and achieving supersonic speeds within another decade. Oh, and we've been to the moon. Each breakthrough, de Grey has written, was "technologically unimaginable to the previous milestone's achievers." Why should aging be any different?

The son of a single mother who styled herself an artist, Aubrey Nicholas David Jasper de Grey attended London's plummy Harrow School on scholarship and studied computer science as an undergraduate at Cambridge. He embarked on a career as a software engineer, but he soon found himself gravitating toward a far messier and more complicated problem: aging.

His interest was more than academic. In 1991, in his late twenties, he married Adelaide Carpenter, a Cambridge professor of genetics who was nineteen years his senior. Under her tutelage, he began educating himself, devouring journal articles about aging science and posting in the online forums of the time. He proved a quick study, publishing his first journal article in 1997, a new theory about the role of mitochondria, the little power plants in all of our cells. That paper later evolved into a book that was impressive enough to earn him a Cambridge PhD, under the university's "special rules" for nonconventional students who had not done graduate studies there (the philosopher Ludwig Wittgenstein earned his Cambridge degree the same way). Armed with that credential, de Grey elbowed his way onto the stage of aging science, wielding a fast-talking debating style, fueled by more than enough arrogance to pull it off. "I am," he assured me over beers, "the most important figure in aging today."

Possibly. For the past decade or more, de Grey has been asking a simple but provocative question: What if we could somehow "cure" aging itself? What if we could defeat it completely, the way we beat smallpox and polio?

In a 2002 manifesto, which he expanded into his 2007 book *Ending Aging*, de Grey outlined a seven-point program by which it would be possible—theoretically—to do just that. His plan, which he calls SENS (for "Strategies for Engineering Negligible Senescence"), would basically engineer the effects of aging out of our very cells...somehow. One way, for example, would be to clear out the "garbage" that accumulates inside our cells, over time. "Your house works fine if you take out the garbage every week, because that's a manageable amount of garbage," he says. "It's only if you don't take it out for a month that you get into a problem."

So all we need to do, to stop or slow these particular effects of aging, is figure out how to empty out our cellular garbage. Somehow.

To call SENS ambitious is an understatement: Another one of its seven pillars entails, in effect, curing cancer. But if it succeeds, de Grey insists, the Vaupel slope will actually get steeper, until eventually we achieve what de Grey calls "longevity escape velocity," where each year we would gain *more* than twelve months of additional lifespan. And thus some of us might theoretically be able to stick around long enough to enjoy whatever it is that people will be doing in 3015 instead of checking Facebook.

Which sounds crazy, maybe even a little scary. Olshansky and twenty-seven other eminent scientists got together in 2005 and published an attack on de Grey and his SENS project, basically saying *Whoa, slow down, cowboy*: "Each one of the specific proposals that comprise the SENS agenda is, at our present stage of ignorance, exceptionally optimistic," reads one of the milder passages. Words like *nonsensical*, *fantasies*, and even *farrago* (a confused mixture, or hodgepodge) were thrown around. "Journalists with papers to sell or air-time to fill too often fall for the idea of a Cambridge scientist who knows how to help us live forever," they huff. They also point out that, by the way, de Grey isn't actually a "Cambridge scientist," as he has never held an academic appointment there, or anywhere else for that matter. (He was employed by the university, but as a computer technician in a genetics lab.)

Perhaps predictably, the attack had the opposite effect: It actually raised de Grey's profile. Rather than ignoring him, the scientists had engaged him. The magazine *Technology Review*, published by MIT, offered a $20,000 prize to anyone who could definitively refute de Grey's theories to a panel of neutral judges. Three groups of scientists took the challenge, but none of their rebuttals was deemed sufficient to win the prize. Another victory for de Grey.

The controversy persists to this day, dividing the field of aging science into two camps: not merely pro- and anti-Aubrey, but those who think we can't do much about aging, beyond maybe tacking on a few healthy years to our hard-won fourscore, and those like Aubrey, who think we might be able to do a lot more, perhaps even reengineering human biology to transcend all its previous limits.

As to when this might actually happen, the jury is most definitely out, and not even de Grey is expecting it to return with a verdict anytime soon. He himself has signed up to be cryonically preserved after his death, à la Austin Powers—that is, immersed in a tank of liquid nitrogen in the hope that, someday, he might be brought back to life somehow. He's not alone: Dozens of others, perhaps hundreds, have signed the forms and paid up to $200,000 for the procedure known as cryonics. The most famous cryonically frozen person, or part of a person, is Ted Williams's head, which now resides in a liquid-nitrogen-cooled tank outside Phoenix. The only hitch is that the technology required to freeze and revive a living creature, even a mouse, does not yet exist. The idea that we might bring back Ted Williams's head, and attach it to a new body—let alone that this would be on the to-do list of some distant future civilization—is perhaps even more far-fetched than the idea that we might engineer our way to immortality. So I didn't take it as a great vote of confidence on de Grey's part.

But in a weird way, although they disagree violently and completely, de Grey and Olshansky are actually saying the same thing, which is that aging is a problem that needs to be solved. Urgently, as a matter of fact. Regardless of what some mainstream scientists think of him, de Grey's big contribution has been to make aging itself—so long accepted as a fact of human existence—a subject of *outrage*.

"Aging is the leading cause of death in the world today," he writes in *Ending Aging*. Run down the list of the leading killers—cardiovascular disease, cancer, diabetes, Alzheimer's, stroke—and you'll find aging as a root cause or major risk factor in ailments that kill a hundred thousand people daily, worldwide. Or, as he likes to put it, "thirty World Trade Centers *every day*."

Yet "old age," hasn't been seen on a death certificate, as a cause of death, since roughly 1952.

De Grey thinks this is a product of society's lingering denial that aging even exists, when in fact it may be our most pressing health problem. Statistics back him up: As this chart shows, one's risk of developing nasty chronic diseases goes up exponentially with age, starting with heart disease in middle age, followed by diabetes, cancer, and eventually Alzheimer's (as well as stroke and respiratory diseases, which also increase dramatically with age).

Add them all up, and it explains why your risk of dying doubles roughly every eight years, a phenomenon first observed by the gloomy nineteenth-century mathematician Benjamin Gompertz in 1825. When we're young, the risk is fairly minimal; there isn't much difference between age twenty-five and, say, thirty-five. But thirty-five to forty-five is a big jump, and by fifty our peers are popping up with breast cancers and colon cancers and high blood pressure and other scary ailments. (Not to mention the beginnings of arthritis, which doesn't kill you but still hurts.)

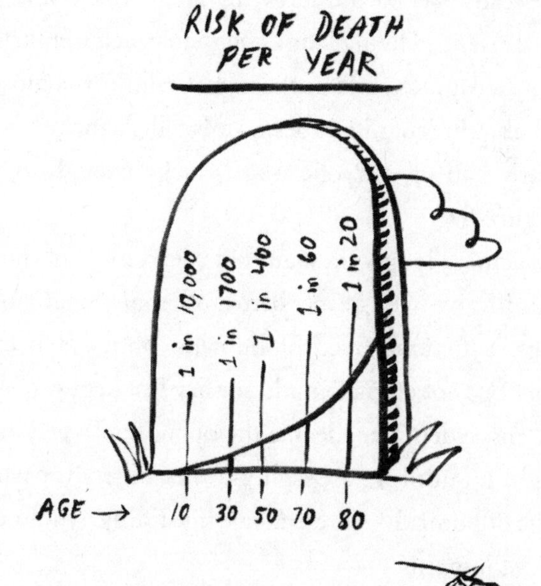

Attacking these diseases one by one, the way Western medicine has done for more than a century, has only helped a little. Over the last four decades, mortality from heart disease has dropped by half, because Grandpa now takes a pill for his blood pressure, another for cholesterol, and he may even have had bypass surgery or a valve replacement, which were not options just a few decades ago. The fatal heart attack at fifty is a thing of the past, yet lifespans have not lengthened nearly as much as scientists had expected. The patient survives one disease, only to fall victim to the next in line.

"We saved people from cardiovascular disease, and then two years later they were dying of cancer or something else," says Nir Barzilai, director of the Institute for Aging Research at Albert Einstein College of Medicine in the Bronx. Indeed, cancer is now the second-leading cause of death behind heart disease, and gaining fast—because people are surviving heart attacks and managing heart disease, and living long enough to get cancer instead. It's like running a gauntlet, and at the end of it lurks the most feared disease of all, the cognitive decay we call Alzheimer's, which affects nearly half of everyone who is lucky enough to make it past age eighty-five.

So if we achieve longer lifespans but spend more of those years in poor health, we will be no better off than Jonathan Swift's Struldbrugs, a fictional race of humans who had been given immortality but not eternal youth, so they just kept getting older, and older, and older. The ideal is the opposite: To stay radiantly healthy right up until we are eighty-five or ninety (or whatever), and then go out quickly—preferably while riding a motorcycle, or maybe BASE jumping.

Unfortunately, the medical research establishment still insists on attempting to tackle the diseases of aging singly: The National Institutes of Health comprises separate, multibillion-dollar institutes for cancer, for diabetes, for heart and circulatory diseases, and so on. This is known as the silo model of medical research, because each disease is isolated from the others. Even now, relatively few mainstream scientists—let alone politicians or policy makers—recognize aging as a crucial underlying risk factor that links all of these problems. The National Cancer Institute gets more than $5.8 billion in funding each year, the National Heart, Lung, and Blood Institute gets $3 billion, and the National Institute of Diabetes and Digestive and Kidney Diseases gets $2 billion. Americans spent more on plastic surgery in 2012 ($11 billion) than we did on government-funded scientific research into the diseases of aging. Meanwhile, the National Institute on Aging gets just $1.1 billion, most of which is earmarked for Alzheimer's disease. The federal government gives just $50 million (or so)—less than the cost of some Manhattan apartments—in annual grants to researchers studying the actual biology of aging.

Which is a shame, because some researchers are beginning to realize that aging itself is the primary risk factor for diabetes, heart disease, cancer, and Alzheimer's—and that something within the aging process itself may link all of them. Each one has a long, invisible beginning, where the disease is developing but we don't have any symptoms. The cellular dysfunctions that lead to Alzheimer's begin decades before we notice any cognitive changes; the same goes for heart disease and diabetes. The thing is, by the time we actually develop the disease, it is almost too late to do much about it. So what if we looked deeper, at whatever it is about aging that makes us prone to these diseases?

In the lab, scientists have made huge strides toward reversing aging, vastly lengthening the lifespans of worms, mice, and flies, often with remarkably simple genetic interventions. In lower animals, and even in mice, knocking out a single gene can nearly double a creature's lifespan, or more. To Aubrey de Grey, that's only a good start. He's proposing a radical reengineering of human biology, one that may or may not ever be possible, with the goal to eliminate or reduce the cellular effects of aging. Nonetheless, it's a provocative idea, and it set up a long-running debate between those who think we can fundamentally alter the aging process itself, and those who feel the best we can do is to live a lot healthier for a little longer.

The American public thinks the latter option sounds like quite enough. A 2013 Pew Research survey found that Americans' "median ideal lifespan"—how long they actually want to live—is ninety years, or about ten years more than we're living now. That's a big jump. But only 8 percent wanted to live past a hundred, perhaps fearing that they, too, will end up like Struldbrugs, teetering and groaning into an artificially prolonged senescence. As Jonathan Swift intuited, immortality has a limited appeal. On the other hand so does dying younger. When public-health thinker and physician Ezekiel Emanuel declared, in an essay in the *Atlantic*, that he wanted to die at age seventy-five—arguing that "over recent decades, increases in longevity seem to have been accompanied by increases in disability"—the backlash was fierce.

Some religious conservatives have lined up against aging research, too. In their view, tampering with aging violates God's will (though they did not make the same argument against antibiotics tampering with God's microbes). The last pope spoke out strongly against life-extension science, and President George W. Bush's official Council on Bioethics—the same folks who effectively banned embryonic stem cell research in the United States—issued a 2003 report declaring that, basically, aging research would result in nothing more than vast numbers of unhappy old people hanging around, getting sicker and spending everyone else's money and making grumpy remarks at holiday dinners.

They were of course dead wrong: Surveys have actually found that in general, older people are far *happier* than their middle-aged kids and grandkids. More to the point, the council's report and the public's skepticism reflects a common fear that Emanuel shared: that longer life will equate to longer *unhealthy* life. The fear is not entirely unfounded: Who would want to spend their extra ten or fifteen years in a nursing home?

The scientists you'll meet in *Spring Chicken* see a very different, and much happier, future of aging. They feel as if they're on the verge of major breakthroughs in our understanding of the aging process, and how we might even begin to modify its course in a way that could enable most people reading this book to live longer, healthier lives—more like my grandfather, and less like my great-uncle.

Even Jay Olshansky agrees, despite his reputation as a pessimist. "I think we're close to a breakthrough, the impact of which will rival the discovery of penicillin," he declared unexpectedly, during our Chicago sausage binge. This breakthrough, probably a medication of some sort, most likely affecting metabolism, would enable most people to delay the most debilitating diseases of aging, at least for a while. Moreover, it would not only transform health care, but the economy, producing what he calls a "Longevity Dividend" in the trillions of dollars. "But we're not there yet," he cautioned.

Which creates a bit of a dilemma. Do you try to hang on for the longevity pill, if it ever comes, and keep yourself together as much as possible? Or should we go out for more hot dogs? (Or cigarettes, even—let's dream big.)

Obviously not—nobody wants to be the last one to miss out on the Fountain of Youth, the cure for aging and the defeat of death for which humankind has yearned for literally thousands of years. But whatever it is, it had better come quickly, because many of the people who are trying hardest to beat aging—people Emanuel called "the American immortals"—are really only making things worse for themselves.

Chapter 3

THE FOUNTAIN OF YOUTHINESS

Nothing makes you look older than attempting to look young.

—Karl Lagerfeld

One wonders what Professor Brown-Séquard might have thought of Suzanne Somers. The hot-pantsed blonde who played Chrissy on *Three's Company* during the late 1970s and early 1980s has evolved into a popular health guru and author of more than twenty books, many of them detailing her own, often elaborate battle with the demons of age. "I am my own experiment," she told a rapt audience of doctors and others at the twentieth congress of the American Academy of Anti-Aging Medicine (A4M) in Orlando, in May 2012.

And how: Every morning, as she revealed to Oprah in a widely viewed 2009 interview, she chokes down no less than forty different supplements, followed by a shot of pure estrogen, administered directly into her vagina. On top of all that, she also takes a daily injection of human growth hormone, which she claims keeps her feeling young. Dinner is served with another twenty pills, which likely resemble the "Restore Sleep Renew," "Bone Renew," and

"Sexy Leg Renew" formulas she sells on her website. That's nothing compared with the two hundred supplements ingested daily by her mentor Ray Kurzweil, the futurist and inventor, who has said he plans to live long enough to see the Singularity, the magical moment when the human brain will (somehow) able to be uploaded to a computer—but it's a pretty good start.

Somers was the keynote speaker at the biannual A4M conference, and she remains the best-known celebrity champion of anti-aging medicine, which is one of the fastest-growing areas of medical practice in the United States. She herself had recently turned sixty-five, but from my seat, in about the fifteenth row, she looked stunning, from her backlit blond mane to her toned shoulders to her blazing crescent smile that made you think that anything was possible.

A few paces behind and to the right of her well-maintained derriere stood Drs. Ronald Rothenberg and Robert Goldman, two of the leading figures in the A4M, which Goldman, a competitive weight lifter, osteopath, and graduate of Belize Medical College, cofounded in 1993. Back then, the organization's annual meetings drew a handful of renegade doctors and longevity enthusiasts, sitting on folding chairs under a tent; this morning, more than two thousand medical professionals packed the ballroom of the Marriott World Center resort and conference center in Orlando, and there would be an even bigger meeting in December in Las Vegas. Worldwide, the A4M claims more than twenty thousand members.

Not bad, for a specialty that barely existed twenty years ago—and that, in fact, still isn't actually recognized by the American Board of Medical Specialties, the arbiter of such things.

Anti-aging medicine remains extremely controversial in the medical profession. A few years ago, Olshansky and his colleague Thomas Perls awarded Goldman and his A4M cofounder (and fellow Belize Medical College graduate), Dr. Ronald Klatz, a mock "Silver Fleece" at a conference in Australia. Klatz and Goldman responded by suing Olshansky and Perls for $150 million, but the case was eventually dropped. "There's no such thing as 'anti-aging medicine,'" Olshansky had insisted, over hot dogs. The secret to aging, in his view, is that there is no secret. "There is no drug, hormone, or supplement, or cream that has been shown to reverse aging, period."

Many A4M attendees clearly chose to believe differently. In addition to Somers's keynote, practitioners could attend lectures on how to prescribe testosterone to aging men (led by the sixty-seven-year-old Rothenberg, who proudly admits pumping himself up to the T levels of a twenty-year-old, so he can keep surfing). Also popular were business seminars that explained how to convert to an all-cash practice. Anti-aging treatments are rarely covered by insurance, and needless to say, Obamacare was a dirty word. In the expo hall, adjacent to the ballroom, the aisles were prowled by statuesque forty-year-old women who turned out, on questioning, to be more like sixty. They were busy perusing the wares, from a scale that told you what to eat to a $6,000 hyper-oxygenated sleeping pod. Olshansky had warned me to beware of "anyone who is selling something," and almost everyone there seemed to be peddling some kind of supplement, special diet, hormone regimen, fancy test, or gadget that would beat back the relentless advance of time. My favorite was an herbal supplement, supposedly derived from Chinese medicine, called Virgin Again.

"We've got a long history in this business," sighed Olshansky, "and it's mostly sordid."

Somers was selling books, mostly. Unlike most Hollywood types, who keep quiet about their struggles with age, Somers has been extremely public about hers, documenting the battle in a series of literary opuses, starting with 2004's *The Sexy Years*. Much of her advice is basic common sense (eat fresh vegetables, get some exercise, sleep well, and manage your stress), and it was hard to argue with her critique of modern American medicine, which keeps older women "all pilled up," in her words. But it all appeared to hinge on those replacement hormones. "I feel so great, I'm loving my life, I have a sex drive," she told the A4M crowd, laughing girlishly. "It's so great! My friends, none of 'em are having sex, none of 'em! And you can see it."

She and her eighty-something husband get busy *twice a day*, as she informed a squirming Sean Hannity on Fox News.

Professor Brown-Séquard would have been blown away, and not by Somers's red-sheathed curves; he didn't care much for sex, subscribing to the common nineteenth-century belief that it sapped one's vitality. But his intuition that the testicles produce some substance that gives males their vitality was proven correct when scientists in Nazi Germany identified the hormone testosterone in 1935 (and won the Nobel Prize for it a year later). Adolf Hitler himself reportedly combated tyrant fatigue by injecting an extract of bulls' testicles called Orchikrin, which was remarkably similar to Séquard's elixir. (He didn't like the side effects, so he soon quit.)

Brown-Séquard had also speculated that there had to be an equivalent for women, and he was right about that, too: estrogen, also identified in Germany in the '30s. Millions of women have used estrogen-replacement therapy, starting about fifteen minutes after the FDA approved its use in 1941, to beat back what Somers calls the Seven Dwarfs of menopause—"*Itchy, Bitchy, Sweaty, Sleepy, Bloated, Forgetful, and All-Dried-Up.*"

Whether they knew it or not, most of the doctors in his hall owed a huge debt to the brave French pioneer, the first modern physician to attempt to combat aging with medicine. But today, Ms. Somers was soaking up all the glory. For a decade, she has been the A4M's leading celebrity spokesperson, the one putting out the hormonally recharged message to the public. Her books have sold more than ten million copies, and each time a new one comes out, she's booked all over the cable dial. The journalists used to give her a hard time for proselytizing on behalf of renegade cancer doctors like Stanislaw Burzynski and Richard Gonzalez, both of whom have long, controversial pasts. (Burzynski has run afoul of both the FDA and the Texas Medical Board, while Gonzalez treats pancreatic cancer patients with a bizarre nutritional regimen centered around twice-a-day coffee enemas.) In 2009, *Newsweek* lambasted her and Oprah for their "Crazy Talk," as the cover headline put it. Now the media mostly lay off, and Somers keeps on keeping on. At the A4M meeting she announced that she had just signed another three-book deal with her publisher. "They can't get rid of me!" she crowed.

But while the publishing industry loves her and her Oprah-esque sales, many experts have a problem with the aggressive hormone treatments that Somers advocates, cranking her estrogen and testosterone levels up to those of a thirty-year-old in her sexual prime (say, Chrissy in *Three's Company*). After *Ageless* came out in 2006, seven prominent women's health experts—three of whom she had quoted in the book—wrote an open letter criticizing her for pushing the "Wiley Protocol," an intensive hormone regimen devised by a writer and actress whose medical credentials consisted of a bachelor's degree in anthropology. "Is it healthy to be dosed like a thirty-year-old when you're sixty?" asks Dr. JoAnn Pinkerton, director of the Midlife Health Center at the University of Virginia Health System. "Show me the evidence."

Hormone therapy is an attempt to solve one of the most obvious problems of aging: We sag. We lose our juice. Men become less manly, women less womanly. Estrogen is a wonderful substance that keeps female bodies fertile, their skin thick and smooth; testosterone builds muscle and gives men the confidence they crave. Both hormones decline in middle age—gradually for men, precipitously for women. Without them, we get dumpy, lumpy, and sometimes Grumpy. So putting them back should solve the problem, right?

You'd think, but in fact the evidence now suggests that solving these problems may create other, worse ones. Hormone therapy for menopausal women was extremely popular and heavily promoted by the pharmaceutical industry until 2002, when the massive Women's Health Initiative (WHI) study was stopped because women on estrogen replacement were getting breast cancer—as well as heart disease, blood clots, stroke, and even dementia—at higher-than-expected rates. If anything, asserts Nir Barzilai, the hormone was effectively *accelerating* their aging.

Overnight, sales of the most popular hormone-replacement drugs collapsed, as did the share price of Wyeth, their manufacturer. Another thing that tanked was breast-cancer rates, which plunged nearly 9 percent in the two years after the WHI study was stopped. Many scientists think those two things are related: Fewer women on hormone-replacement therapy led to fewer cases of breast cancer. But some women remained desperate for a fix, understandably, and so Somers stepped into the breach two years later with *The Sexy Years*, in which she offered an alternative: so-called bioidentical hormones, which are chemically identical to the hormones produced by the female body. She claimed these were safer than Wyeth's products, which were synthesized from pregnant mares' urine (hence the name, Premarin). Somers has been pushing bioidentical hormones relentlessly ever since, claiming they are less dangerous and more "natural."

"Bioidenticals are a natural substance and cannot be patented," she writes in *Ageless*. "Therefore, there is no money to be made from selling the best solution for menopausal women."

Actually, not true: bioidentical hormones are custom-mixed by compounding pharmacies, which presumably earn a profit by selling them. (Also, they're rarely covered by insurance, so they cost women more.) They generally contain not only estrogen, but balancing hormones such as progesterone and sometimes testosterone, among others. The upside is that this allows a doctor to prescribe a precise mix for each patient. The downside has to do with how they're made, and what might be in them: Compounding pharmacies are lightly regulated, unlike pharmaceutical manufacturers who are subject to strict FDA rules. Studies of compounded drugs have found that the actual dosages can vary enormously, and in 2012, contamination at a compounding pharmacy in Framingham, Massachusetts, caused an outbreak of fungal meningitis that killed sixty-four people and sickened more than seven hundred. The pharmacist was indicted in September 2014.

"In my practice I end up picking up the pieces from people who believed [Somers]," says Dr. Nanette Santoro, an expert on menopause at the University of Colorado in Denver. "I've had a woman bring in her hair in a Baggie because a compounding pharmacy screwed up her prescription."

Women don't have to go to compounders: There are now several FDA-approved bioidentical hormone treatments on the market, a fact Somers never mentions. FDA-approved means a drug has been tested for safety, efficacy, and dosing and absorption; just as important, your doctor knows how much you are really getting. But unlike the compounded drugs, the FDA-approved hormones are required to carry a scary "black box" warning label.

By her own account, Somers has been using hormone therapy for at least twenty years, even after surviving her own brush with breast cancer. In her latest book, *I'm Too Young For This!*, she recommends hormone therapy for still-younger women, beginning as early as their late thirties. So she essentially endorses hormone use for more than half a woman's life—despite evidence that such long-term use is clearly unsafe. In fact, says UVA's Pinkerton, the evidence shows that hormone use is safe only for short periods of three to five years—and that after age sixty, the risk rises dramatically. "The current theory is there is a critical window, where if women [are] given hormones for a short period around the time of menopause, they may have a benefit for heart and brain," says Pinkerton. "But once you have plaques in your arteries, or aging neurons, giving estrogen may accelerate those problems."

"Women are unusually susceptible, because they have this abrupt change in their lives," Santoro observes. "Other than the wisdom it brings, everything else is a problem."

Just as women have been catching up to men in all areas of life (except for equal pay, domestic abuse, and reproductive rights), men can now claim equality in one important respect: Now we get to go through menopause, too. It's called *andropause*, and it refers to the long-term decline in testosterone that becomes noticeable around age forty. It's nothing like menopause, obviously—women's hormones nose-dive off a cliff, while male testosterone levels descend on a gentle glide path—but nevertheless, testosterone replacement for men has become almost as big as estrogen replacement was before the WHI study ruined the party. According to some estimates, "Low T" therapy is now a $2 billion business, and could nearly double by 2019.

Certainly, Brown-Séquard would be amazed by the progress we've made. Instead of his nasty old bull's-testicle brew, aging men can now deploy a convenient underarm gel, which is advertised during every NFL game (but which their wives mustn't touch under any circumstances, according to the warning label). In the ads, pudgy sad middle-aged men are transformed into confident, smiling satyrs, although the reality isn't quite so simple. In smaller, short-term studies, testosterone has been shown to increase muscle mass, improve mental sharpness and overall well-being (and enhance libido, although that part is surprisingly controversial among scientists). There is even a study that purports to show that testosterone administration makes men less prone to lying. But there is little data about its long-term safety. The most common concern, that it feeds prostate cancer, is not supported by evidence; the notion actually stems from a single case, reported in 1941. (In recent studies, doctors tried to use testosterone to prevent or treat prostate issues, but with limited success.)

There are other serious issues, however. A 2010 study of testosterone in men with heart problems had to be stopped because it seemed to increase the risk of cardiac events. Another study also reported an increase in stroke risk. Other studies funded by manufacturers showed little to no increased risk, but as the authors of one review put it, tartly, "The effects of testosterone on cardiovascular-related events varied with the source of funding."

Unfortunately, there has been no large-scale clinical trial, similar to the Women's Health study, looking at the safety of testosterone replacement in older men. But the FDA has become sufficiently concerned that in March 2015 it required a severe warning label to be placed on all testosterone-replacement products, warning of increased risk of cardiovascular events and stroke. Meanwhile doctors are writing prescriptions even faster, driven by advertising and promotion rather than actual medical need, according to the Australian testosterone researcher David Handelsman. Handelsman says testosterone is being drastically overprescribed, and dismisses the very concept of andropause itself as a "false analogy" to the very real, very severe life change that is menopause.

It seems doubtful that any expert opinion, study result, or warning label would deter Somers or her A4M followers one bit. On stage, she radiated confidence, mesmerizing the audience with her passionate delivery. She was most excited about the advent of "nanoparticles," tiny little entities that will travel in our blood, collecting information and dispensing treatment to us from within. Ray Kurzweil had told her about them.

"I can't believe how great the view is from sixty-five!" she gushed. "Ray Kurzweil asked me how long I thought I would live, and I said, 'Honestly, Ray, I can wrap my arms around 110. Honestly. With the way I feel. With strength.' "

As I sat there in the audience, admiring the giddy-blond sex symbol of my teenage years, I tried to imagine what a 110-year-old Suzanne Somers might look like. Answer: probably pretty good, for 110. She certainly wasn't going to go down without a fight. But I had doubts as to whether her chosen methods would get her there.

My curiosity was piqued by something that she only glancingly mentioned, but which underlay much of what went on at the A4M conference: human growth hormone, or HGH, prescribed by many if not most of the anti-aging doctors in attendance. The very name itself is almost talismanic: *human growth hormone.* Who wouldn't want that?

Not only does Somers take it daily, but one of the foundational texts of the anti-aging movement is a 1997 book called *Grow Young with HGH: The Amazing Medically Proven Plan to Reverse Aging*, written by A4M cofounder Ronald Klatz. The book touted the amazing powers of human growth hormone, a potent drug mainly used only by stunted children and aggressive body builders. Once harvested from the pituitary glands of cadavers, HGH was first synthesized in 1985, at the beginning of the biotech revolution. Since then, the HGH market has taken off. Based on my chats with several doctors present, it was clear that HGH and hormone replacement injections were a major profit driver for their practices. The shots, and the attendant blood work, can cost patients upward of $2,000 a month. "It's changed my life," said one male physician from Florida, a former family doctor who says his income tripled when he started selling hormones to aging Boomers.

But what about the "medically proven" part? Can HGH really "reverse aging," as Klatz's book claims? Is it really the magic key? Certainly, cheating athletes from Lance Armstrong to Alex Rodriguez seemed to think it did *something* for them. And Somers is certainly not the only Hollywood type to partake. Sylvester Stallone was busted with a suitcase full of the stuff a few years ago, for example, and the late model/actress Anna Nicole Smith was also an HGH user when she died in 2007 at age 39.

"I'll tell you why I took HGH in the first place," a sixtyish man identified as a "longtime Hollywood filmmaker" told *Vanity Fair* in 2012: "I love fucking."

Better advertising than that cannot be bought. Between 2005 and 2011, U.S. growth-hormone sales soared 69 percent, to $1.4 billion, according to an investigation by the Associated Press. Nobody knows how much more is imported illegally from China, India, and Mexico. The big numbers belie the fact that HGH is extremely difficult to prescribe legally. The hormone is FDA-approved only for a narrow set of fairly rare conditions, including "short stature" in children and wasting in AIDS patients. Off-label prescribing, for conditions not specified in the drug's FDA approval, is officially frowned upon; indeed, HGH is governed by stricter rules than pharmaceutical-grade cocaine, largely to prevent abuse by athletes. The anti-aging docs get around that by diagnosing their patients with something called "adult growth-hormone deficiency", a vaguely defined condition that could be applied to nearly all older people, since growth hormone declines naturally from about age twenty onward. Pfizer was fined almost $35 million in 2007 for illegally promoting off-label use of HGH, but that was a drop in the bucket compared with its sales. According to the AP, "At least half of [2011's] sales likely went to patients not legally allowed to get the drug."

More recently, the federal government has begun to crack down, launching stings on anti-aging clinics like the one in Miami that was frequented by Rodriguez. Perhaps not surprisingly, then, the A4M folks were reluctant to talk about HGH on the record. Actually, "reluctant" is putting it mildly: Their PR woman casually mentioned that the organization might consider suing me, the way they did to Jay Olshansky. The heat was on. So I flew to Las Vegas to meet with the most outspoken growth-hormone user I could find: the famous Dr. Life.

If you've ever flipped through the back pages of an airline magazine, you'll recognize Dr. Life: He's the balding, smiling grandpa with the implausibly muscled torso. Amazingly, Dr. Life is his real name: Aged seventy-five, but built like a twenty-five-year-old Chippendale's dancer, Dr. Jeffry Life is the public face of a Las Vegas–based company called Cenegenics, which operates more than twenty anti-aging clinics nationwide.

In person, Jeff Life is down-to-earth, friendly, and every bit as pumped as his famous portrait, which hangs in his office, in Cenegenics' white marble palace of a headquarters building, about twenty minutes from the Las Vegas Strip. His celebrated biceps bulge under the sleeves of a simple black T-shirt as he leans back in his chair and puts his feet up on the desk, which is perfectly clean. The shelves are occupied by copies of his two best-selling books, *The Life Plan* and *Mastering The Life Plan.* This does not look like an office where a lot of actual work happens; it's good to be Dr. Life.

One other thing I notice is that while his torso says *gay porn star*, his face and hairline say *Larry David.* He has clearly not had any plastic surgery or even hair transplants. He looks his age, except for the huge pecs. This makes me want to like him, but for different reasons than I wanted to like Suzanne Somers.

On the other side of his office hangs another large framed photo, of a dumpy, beer-bellied guy, slumped on a sailboat on a muddy lake. That, too, is Dr. Life. Or rather, it was. When the picture was taken, he was a fifty-seven-year-old family doctor in northeastern Pennsylvania, with a lousy marriage, a flabby gut, and a boozy thirst. He didn't know it yet, but he was well on his way to Type 2 diabetes, and he already had coronary artery disease. A couple of years later, as he tells it, a patient happened to leave a copy of a bodybuilding magazine in his waiting room, and he took it home and read it cover-to-cover.

A few issues later, he decided to enter a contest sponsored by the magazine, offering a prize to men and women who most dramatically remade their bodies by working out with weights. His girlfriend, Annie, now his wife, snapped a "before" photo in his paneled basement. Then he joined a gym, and hired a trainer and nutritionist, and pumped iron like mad. Twelve weeks later, he sent in his "after" picture, which is also framed on the wall. The beer-bellied putz on the sailboat had been transformed into a basement Schwarzenegger, even more ripped than in the famous ads. He won the "Body-for-Life" contest, but he had also changed his own life. In twelve weeks, he had become practically a different person.

"This is incredible," I say.

He nods. "I tell people, if I can do this, anybody can do this," he says. "I am not special. In fact, I have inherited bad genes, and I have trumped my genes. I am actually a fat person in a lean body."

But—as you might have guessed by now—exercise and eating right are not his only secrets. He kept training intensively, but by the time he reached his mid-sixties, he felt he was losing ground. "I was still going to the gym and trying to eat good, but I was gaining back my belly, and I was losing muscle mass and strength," he says. In 2003, he attended a Cenegenics conference in Las Vegas, where he learned about the importance of "healthy hormone levels." He stayed an extra day to undergo Cenegenics' $3,000 initial patient evaluation, which included blood tests, fitness tests, and various body scans to determine his body composition and to detect cancers.

The blood tests showed he was at the lower end of the normal range for testosterone and growth hormone, even for his age. His new Cenegenics doctor started him on a regime that maintained his intensive weight-lifting workouts, but added twice-weekly testosterone shots, plus a daily injection of HGH. "I started to feel better within two weeks," he says.

Within a year, he had moved to Las Vegas, where he soon became Cenegenics' poster boy and a protégé of its founder, Dr. Alan Mintz, a former radiologist and competitive weight lifter who was then in his early sixties. Mintz himself was an outspoken advocate for and enthusiastic user of human growth hormone, as well as other things. According to one story, Mintz had once run the New York Marathon while carrying syringes loaded with painkillers (because of an injured knee).

The other pillar of Dr. Life's program, is, of course, a rigorous-bordering-on-brutal workout program, alternating between intense strength training and equally intense cardio—riding a Lifecycle (no relation) while watching DVDs of action shows like *Breaking Bad*. He's got two stents, because of the coronary disease, but hasn't let that stop him. He eats a sensible diet—tonight's dinner will consist of a skinless chicken breast, brown rice, and broccoli—and he gave up drinking years ago. "You can't drink alcohol and look like I do," he acknowledges. (Note to self.)

But he still insists the injections are essential. "I realized that the missing link was hormones," he says. Without growth hormone, "I am absolutely, 100 percent sure that I would not look like I do, feel like I do, act like I do, or think like I do."

Which sounds fantastic; sign me up to look like Dr. Life when I'm seventy-five. But as I dug into the actual scientific research on human growth hormone—and the recent history of Cenegenics itself—I found that the controversial drug is not quite the Fountain of Youth that many people believe it is. Far from it; in fact, many scientists believe that it may even help *accelerate* the aging process.

Amazingly, much of the hype for HGH is based on a single, small study that was published in 1990—and which has since been disowned by the journal where it appeared. In the study, a researcher at the Medical College of Wisconsin named Daniel Rudman gave HGH injections to a dozen patients over sixty with below-average levels of growth hormone for their age. After six months of thrice-weekly injections, plus a modest workout program, Rudman found that the men had gained more than ten pounds of "lean body mass" (aka muscle), while losing nearly eight pounds of their fat.

Up to that point, recombinant HGH had been an obscure little drug, known mainly to pediatricians who treated children with growth deficiencies. After the study appeared, the adult-growth-hormone market exploded almost overnight. Finally, a drug had been "scientifically proven" to increase muscle mass while making fat disappear. Ron Klatz gratefully dedicated *Grow Young with HGH* to Rudman—but Rudman himself, who died in 1994, was horrified by the way his work was misused. His tiny study was cited in so many ads by mail-order pharmacies and sketchy anti-aging clinics that in 2003, the *New England Journal of Medicine* took the unusual step of denouncing its own publication, in a strongly worded editorial: "Although the findings of the study were biologically interesting, the duration of treatment was so short that side effects were unlikely to have emerged, and it was clear that the results were not sufficient to serve as a basis for treatment recommendations."

Too late. The horse had already fled the barn, and it was jacked up on HGH. But as I found on further research, HGH may not exactly live up to its hype. While it does appear to increase muscle mass and vaporize body fat, it does not appear to increase actual muscle *strength*. Lifting weights (and taking testosterone) do improve strength; also, weight training and vigorous exercise like sprinting have been shown to increase growth-hormone levels naturally. So does deep sleep. Which made me wonder whether Dr. Life really needed the stuff: He eats well, doesn't drink, and exercises vigorously, doing a mix of strength training and aerobic conditioning. Perfect. Does he really require help?

He insists that he does: Growth-hormone injections are the only way a guy his age could look like... Dr. Life. But unfortunately, it does other things, too. Growth hormone has been linked to a long list of side effects that include edema, joint pain (including carpal tunnel syndrome), and a much-increased risk for glucose intolerance and even diabetes in older patients, particularly men. As for the long-term effects, it's hard to say. Unlike with estrogen, there are no large, long-term clinical trials of HGH use in older adults, in part because it's technically illegal to use it for anti-aging purposes, but also because neither the pharmaceutical companies who make the stuff nor the doctors who prescribe it have shown any enthusiasm for such a study. So their millions of patients are essentially experimenting on themselves, just like Suzanne Somers (and Brown-Séquard).

For a while, I considered doing the same thing, signing up for testosterone-replacement therapy and possibly even taking some human growth hormone, just to see how it would feel. I even had myself tested, but found that my testosterone levels—not to brag—were already pretty high. Growth hormone remained intriguing, however. But it didn't take long for me to discover some compelling lab science that explained why taking more of it might not be such a great idea. Lab science, and also the movie *The Princess Bride*.

Let's start in the lab, with this fun fact: The longest-lived mice ever observed in a laboratory actually had *no* growth-hormone receptors in their cells. They had been genetically modified to be immune to the juice. Wrap your head around that for a minute. No growth hormone, longer life. This phenomenon was first discovered by a postdoctoral student named Holly Brown-Borg, who was sorting through a batch of mice, picking out older animals for an aging study she had planned, when she noticed that many more of the older mice were of a type called "Ames Dwarf," which had a strange mutation that knocked out their growth-hormone receptors. She took a census and found that the dwarves seemed to live longer than the others. "It was like wow, maybe growth hormone has something to do with it," says Brown-Borg, now a professor at the University of North Dakota.

Long story short: She and her mentor, Andrzej Bartke of the University of Southern Illinois, found that growth hormone and longevity were actually inversely related. Some strains of growth-hormone "knockout" mice have lived nearly twice as long as normal mice, or almost five years. They are much less likely to develop the diseases of old age, like cancer, which means they really are aging more slowly. Meanwhile, mice that have been bred to produce extra growth hormone tend to live only about half as long. All of which suggests that excess growth hormone might be unhealthy. So why shoot up with more of it?

"I wouldn't take growth hormone as an older person," says Brown-Borg. "It accelerates aging, if anything, in the long term. I don't understand why people would even think of taking it."

I can: Users seem to think HGH makes them feel younger, at least for a while. But so would getting a pair of nipple rings. In the long run, neither choice is particularly wise.

For more evidence, you need look no farther than the local dog park. Small dogs like Chihuahuas can live for fifteen years or more, while Great Danes rarely live longer than seven or eight. The reason for their vast difference in size comes down to a single gene for insulin-like growth factor 1 (IGF-1), the messenger that tells our cells to grow and divide (and which works hand in hand with growth hormone). The bigger dogs produce more IGF-1, while the small dogs have had it winnowed out of them, through hundreds of years of selective breeding. Could that also explain why the smaller dogs, with less growth hormone and less IGF-1, almost always live longer than bigger dogs? And does the same hold true for humans?

Certainly, excess growth hormone was not a good thing for André the Giant, the beloved pro-wrestler-turned-actor. He suffered from a rare condition called acromegaly, which is caused by a benign tumor on the pituitary gland, causing it to produce too much growth hormone. (The motivational speaker Anthony Robbins has the same condition.) He towered well over seven feet tall and weighed more than five hundred pounds in his prime; talk about "big-boned." He lived like a giant, too: When he wasn't knocking the pancake makeup off Hulk Hogan or walking away with *The Princess Bride*, he could reputedly put away more than a hundred beers at a sitting. That may have helped hasten his untimely death in 1993 at just forty-six, but people with acromegaly often die fairly young.

One reason may have to do with cancer. Excess growth hormone is known to stimulate the proliferation of cancer cells, although it's not known whether growth-hormone users are at greater risk for contracting cancer, since the proper studies have not been done. There are some troubling cases. In 2003, a fifty-six-year-old California woman named Hanneke Hops came to Cenegenics for anti-aging treatments, because she wanted to keep enjoying her active lifestyle, running marathons and riding horseback. Like many if not most other Cenegenics patients, she was given growth-hormone shots, which can cost from $1,000 to $2,000 a month (for the shots, the required tests, and office visits). She lost sixteen pounds right away. "It makes me feel good," she told the *San Francisco Chronicle.* But her treatments did not last long, because six months later she was dead, her liver riddled with malignant tumors. Her family claimed the growth-hormone treatments caused or accelerated her cancer, but Dr. Mintz insisted they had not.

A few years later, Mintz was profiled by *60 Minutes*, which was investigating the controversy around anti-aging medicine—but before the story could air, Mintz, too, had died under mysterious circumstances. At first, the company had said he had suffered a heart attack while lifting weights, but it later emerged that he had succumbed to complications from a brain biopsy performed by doctors looking for evidence of cancer. The biopsy had come out clean, the company insisted. But no autopsy report has ever been made public, so the true cause of his death in 2007, at the not-so-old age of sixty-nine, may never be known.

What is clear is that aging is a lot more tricky than most people realize—and that simply putting something back, like growth hormone or whatever supplement happens to be trendy at the moment, isn't going to solve the problem, any more than Séquard's Elixir did. There's no quick fix, despite what the many hucksters at the A4M meeting were offering: the hormone-replacement regimens, but also the brain-wave tuners, the hyperoxygenated sleeping chambers, and the seemingly endless array of supplements, all promising to clean up this or that unfortunate side effect of growing old.

"Imagine you have a symphony written by Mozart," says Valter Longo, a professor of biology at USC and a leading researcher on aging. "Taking growth hormone or a supplement or whatever is like someone going to the cello player and saying, 'Can you just make it a lot louder?' Chances are, it will screw things up."

* * *

Still, Brown-Séquard was on to something important—and so, in their own deluded ways, are Suzanne Somers and Dr. Life. An older body is fundamentally different inside from a young one.

In the early 1970s, a German-born scientist named Frederick Ludwig at the University of California–Irvine showed how important this is, in a novel and radical experiment: He sliced open the right side of a three-month-old rat, and the left side of another rat that was about eighteen months old, the equivalent of a sixty-year-old person. He then stitched the two animals together, from shoulder to flank, like Siamese twins. He did this, again and again, until he had 235 pairs in different combinations: old with young, old with old, young with young.

The technique was called parabiosis, and if it sounds like something from Brown-Séquard's day, that's because it is: It dates from the 1860s, when a French physician named Paul Bert first joined two albino rats together. He discovered that the two animals' circulatory systems joined up, so the same blood flowed through both bodies. Since then parabiosis had been used to study the immune system, kidney function, cancer, and the effects of radiation. Ludwig's question was elegantly simple: What would happen if you hooked up an old animal to a source of young blood?

The idea was not his alone: The notion of "heterochronic" parabiosis, pairing old and young animals, had been suggested several years earlier by Alex Comfort, an early British gerontologist who intuited that there was something very special about youth—how young people resist stress, injury, and disease so much more powerfully than their parents. "If we kept throughout life the same resistance to stress, injury, and disease that we had at the age of ten," he wrote, "about one-half of us here today might expect to survive in 700 years' time."

That isn't a typo: *700 years*. It's the stuff of Aubrey de Grey's dreams. All you'd have to do is stay the same, biologically, as when you were ten. That's the trick.

Comfort suspected that there was something in young bodies that gives them their wonderful, youthful powers of resistance and regeneration, and while this thing is extremely potent, it is also ephemeral, seeming to vanish within a decade or so. But Comfort was too busy to do the experiments himself, because at the age of sixty-something he was busy putting the finishing touches on his most popular work, a handy little illustrated instruction manual called *The Joy of Sex*. So the heavy lifting fell to Ludwig and his lab assistants, who worked for weeks stitching animals together.

Parabiosis is admittedly a bit creepy; it's actually banned in some countries. But it sounds a lot worse than it really is. I witnessed a parabiosis surgery on two rats, in a lab at Albert Einstein College of Medicine in the Bronx, and it took about twenty relatively bloodless minutes to cut them open and suture them together. Within a week or so, the incision would heal and their circulatory systems mesh, so that the same blood would be coursing through both bodies. True, more than a third of Ludwig's pairs died, but survival rates are far better now, and they go on to live relatively contented lives, once they sort out which animal will "lead," like in ballroom dancing. Certainly, their new, joined existence is more interesting than sitting around in a cage alone all day long. "When you think about it," one scientist involved in parabiosis experiments half joked, "it's really pretty boring being a lab mouse."

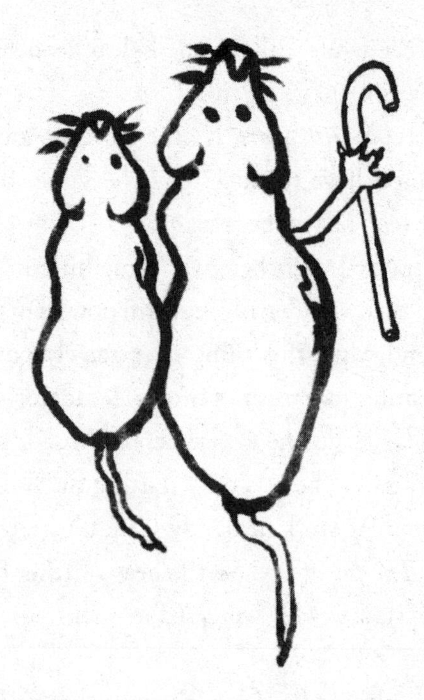

Next, Ludwig and his team did what scientists always do in aging studies: They waited for the animals to die. (Talk about boring.) The results were anything but dull, though. The old rats that had been paired with young animals went on to live an amazingly long time, between four and five months longer than rats joined to partners of similar age. They even survived slightly longer than Ludwig's sixty-five single control rats, who had not been subjected to the traumatic surgery. Given that lab rats normally live just a little more than two years, this was like extending human lifespan from roughly eighty to nearly a hundred years.

In other words, Ludwig found that youth is contagious. But why?

The best explanation he could offer was that the older animals may have been protected from infection, benefiting from access to the younger immune system. The immune system is certainly important to aging, as we'll see, but was there a more profound reason? Was there something in young blood, some sort of secret youth-giving factor, that helped the older animals live longer?

It's an old, old question. As long ago as the thirteenth century, the alchemist-philosopher Roger Bacon asserted that an old man could be rejuvenated if he inhaled the breath of virgins (male, supposedly). Many an old man (and woman) since has sought the company of youth, sexual or otherwise, perhaps with a similar goal in mind. In the sixteenth century, Sir Francis Bacon (no relation to Roger) performed a blood transfusion from a young dog to an old one, who seemed to him quite rejuvenated. And speaking of young blood, let's not forget the fictional Dracula, who dined on nothing but, and was alleged to be several centuries old—although in real life, the actual fifteenth-century Count Vlad "The Impaler" Dracula only made it to his mid-forties.

For the time being, even Ludwig could not provide any better explanation of what he had observed, and parabiosis soon fell out of fashion once again. But the questions that it had raised were profound: Something in young blood must carry this power to rejuvenate old animals—but what was it? And why do we lose whatever-it-is with age? What changes inside us, so profoundly that it alters the makeup of our very blood?

Or should I quit asking these questions and just go get a blood transfusion from my young niece—who, as of this writing, just happens to be exactly ten years old?

That's what I needed to find out.

Chapter 4

YOURS SINCERELY, WASTING AWAY

The most mind-blowing thing for me every day is the mirror.
I can't believe what I look like.

—Neil Young

I awoke at sunrise to find a strange man sitting by my bed, preparing to strap a rubber mask over my face. Working quickly but gently, he fitted it over my nose and mouth, sending me from a state of groggy half wakefulness into an instant claustrophobic panic. "It's okay," he said soothingly. "Just relax."

"Mmmph," I groaned from behind the mask as my arm flopped spastically.

"Ssshhhh!" he said, capturing my errant arm and pinning it to my side. "Just lie there quietly."

I lay back and tried to think about anything but the tight-fitting mask and my very real suffocation phobia. The man in my room was named Edgar, as in Allan Poe, and he was the night nurse at Baltimore's Harbor Hospital, where I had just spent my third night in a row. I was not here because I was sick, but because I was healthy: I had volunteered for something called the

Baltimore Longitudinal Study of Aging, or BLSA, the world's longest-running study of human aging. Since 1958, government researchers have been monitoring a growing cohort of subjects on their ride down the roller coaster of Time.

The study was the brainchild of the pioneering gerontologist Nathan Shock, who was one of the first American scientists to study aging at all. After graduate school, he had landed at the Baltimore branch of the National Institutes of Health, where he soon realized that, basically, scientists had no idea how people naturally grow old. For one thing, until the mid-twentieth century there had never been all that many old people around. Also, the gerontologists of the time tended only to study people who were *already* aged, or even dead.

Shock's stroke of genius was the "longitudinal" part: Rather than poke and pick at old people, he would start with healthy folks who were not quite so old, and observe what happened to them as they slowly aged. So instead of comparing a hypothetical average seventy-year-old to a hypothetical average forty-year-old, they would track each individual on his or her unique journey into old age. He rounded up a core group of subjects, mostly fellow scientists and physicians in the Baltimore medical community, and performed all kinds of basic tests and measurements on them. He would then watch as they changed over time.

Such a long-term project would be almost inconceivable under today's tight funding deadlines and tenure-track pressures, but Shock's little study has matured nicely. The BLSA is now following more than thirteen hundred subjects between the ages of twenty and 105. His small office has become the National Institute on Aging. Over the years, NIA scientists running the BLSA have compiled a painfully detailed rap sheet of the crimes committed upon the human body by aging. According to one BLSA study I pulled up on my laptop, average VO_2 max—the body's ability to process oxygen while exercising—declines by 10 percent during one's forties, 15 percent in the fifties, and 20 percent in the sixties—and, oh yeah, 30 percent in the seventies. One thing that doesn't decline, however, is subjects' weight, which packs on relentlessly through the decades of middle age. (Thanks, Science.)

Some of the information has actually helped people. One of the BLSA's most important findings was that for men, levels of prostate-specific antigen (PSA), a marker for prostate cancer, don't really matter; what's important is the *rate of change* in PSA. That alone has saved thousands of men from needless painful tests and surgeries. More recently, BLSA data has been used to establish the diagnostic criteria for diabetes, and it has also helped researchers understand the patterns of progression of cardiovascular disease and Alzheimer's. But its big question remains largely unanswered: How do you *measure* aging? Is it possible to determine someone's biological age, versus their chronological age?

The BLSA doesn't take just anybody. I'd had to undergo a thorough screening, with blood tests, a complete physical (minus prostate exam), and many nosy questions about my medical history, all to make sure I was healthy enough for the government to watch me get old and fall apart. If I had any chronic ailments or was taking any medications, including even just ibuprofen, I would have been disqualified. "These," cracked the intake nurse, "are some of the healthiest people you'll ever meet."

For the privilege of joining this elite cadre, I gladly gave up three perfect August days to serve as a government guinea pig. I would spend every waking moment undergoing tests of one kind or another, and even sleeping on the hospital ward, in a room with a beautiful view of Baltimore Harbor. All told, the government would collect six thousand pieces of data on me, and every few years, I would be required to come back to give them more. Although I would be free to drop out anytime, the study would not necessarily even end when I died: A brochure outlined the option of donating my body to the BLSA. In return, I would be receiving—for free—the best, most complete medical evaluation that taxpayer money could buy. And also, in some ways, the weirdest.

Already, over the preceding two days, I had been poked, prodded, stabbed, jabbed, drained, and scanned in every imaginable way. It started early on Tuesday, when a nurse systematically relieved me of about thirty vials' worth of blood. Already queasy from this ordeal—like most men, I do not enjoy watching blood exit my body—I then had to chug a bottle of a sickly-sweet orange-flavored drink, and sit there while she pilfered yet more of my blood, every twenty minutes for two hours, to test how I coped with a massive onslaught of glucose. Or, as they say in Georgia, "breakfast."

Some of my blood was analyzed on the spot, but most would be banked in NIA freezers, for use in future research projects. (Note to future civilizations: Please unearth those vials and clone me.) We then moved on to a series of crazy-making cognitive tests that required me to memorize and recite a shopping list that seemed to have been compiled by an insane person, or possibly my girlfriend: "Squid, cilantro, hacksaw, perfume..."

Everything that could be measured, was. For the first twenty-four hours, I was forbidden to urinate anywhere but into an orange jug, which resided in a cooler in my bathroom. There was more: A prim young woman had taken close-up photos of my tongue, which she had first dyed blue. "This isn't my favorite part," she admitted. Mine either. In another room, a friendly, chubby guy had plastered electrodes to my face and tapped me on the forehead with a rubber mallet, twenty times. This was purportedly to evaluate the nerves that stabilize our vision, like our own internal Steadicam. A nurse trimmed my toenails and saved the clippings in a vial so they could analyze my "microbiome," the community of microbes that dwell inside, on, and around us, and whose importance has only recently been recognized. Further to this end, I learned to my horror, they would also be requiring a sample of my poop. For this, I was given a special collection device that resembled one of those old-time ladies' hats, like they wear on *Downton Abbey*.

I feared only the MRI machine, where I'd be loaded into a narrow, claustrophobic white tube that made horrible screeching noises for more than an hour, as it scanned my brain and then my legs. It wouldn't be all that different from flying coach to Cleveland, really, but still, I hated the idea. After a little coaxing from the technician, whose name was Bree—"I'll be *right here*," she cooed, touching my arm—I gladly let her strap me to the tray and slide me into the machine, which then recorded images of my brain taking an imaginary vacation with Bree to an all-inclusive tropical resort.

My stint in the BLSA was, hands down, the highlight of my research for this book. I won't lie: It was kind of nice to get so much attention for three days, especially for a guy who normally works at home, alone. It was good to be busy. And the tests seemed ridiculously easy (well, except for memorizing that shopping list). It was so much fun that I took to calling it The Blast.

On my first day, for example, I was marched into a hallway, placed in front of a chair, and told to get up and sit down ten times, arms crossed over my chest, while a staff physiologist timed me with a stopwatch, like a football scout. I then had to stand on one foot for 30 seconds, then the other, a test of balance. *Nailed it.* The physio also timed me as I walked up and down in the hallway with an oxygen mask on my face (they were big on oxygen masks). Aced that one, too. The next day, I had to walk some more—this time, in a state-of-the-art "gait analysis" laboratory, where my movements were recorded by the same kinds of high-speed cameras used by Pixar. I successfully placed one foot in front of the other.

The vision exam was a breeze, even when they shone headlights in my face, to simulate oncoming traffic, and I got down to the faint squeaky tones on the hearing test. I also put up a pretty good fight against the grip-strength-o-meter, and the leg-extension machine—just like Brown-Séquard.

But the best part was the way Blast staffers constantly kept telling me how "young" I was, which was true relative to their typical clientele. This alone made it a wonderful experience. The nurses and doctors and physiologists were happy because I only had to be briefed on each test once, and I did everything quickly. I was rocking The Blast.

Only later did I realize what this meant: that I had only begun to experience real aging. I would come back every three years, and every time, most of the things they were measuring would likely get worse, not better: body-fat percentage, bone density, vision, hearing, strength, cardiovascular health, glucose tolerance (the syrupy drink test, simple but extremely important). Oh, and memory, obviously. Soon I would be forgetting to turn on my cell phone. The interview about my urination habits would become more lengthy and awkward. To continue the liquid metaphors, aging was like a river that flowed in only one direction: downhill.

All this began to dawn on me about halfway through the quarter-mile walk test, twenty laps up and down the corridor, as fast as I could go. Even though I walked "briskly," as instructed, I barely broke a sweat. But then it hit me: *Someday, this will be hard.* Getting out of a chair will become a painful, humiliating ordeal. Just walking a few blocks will seem like a monumental task. And not long after that, I will die. This was actually one of the most important tests, in that regard. Thanks to Blast data, researchers now know that natural walking speed is one of the most accurate predictors of mortality that we have. The slower you walk, statistically speaking, the sooner you are likely to check out.

I started to get depressed. Yes, I was still in pretty good shape. For now. But aging had barely affected me yet, and frankly, I was being kind of a crybaby about the whole thing. My long, inevitable decline, which is what The Blast would really be measuring, had only just begun. This was as good as it was going to get. Even my height would change—I'd probably never be this tall again, as the study has documented a long, steady decline in stature (due to water loss in the discs between our vertebrae, if you must know). The study wouldn't end until I died, and not even then if I signed the autopsy consent form.

That night, I went out and devoured half a dozen Chesapeake blue crabs, among the highest-cholesterol foods there is, washed down with several beers. Because why not?

A few weeks after I got home, I phoned Dr. Luigi Ferrucci, who then headed The Blast. When he had taken over in 2002, the study had been languishing, as it had since Nathan Shock died in 1989. It was considered unfashionable to study aging in actual human beings, an exercise some molecular biologists derided as "wrinkle counting." Instead, the National Institute on Aging had spent millions of dollars on mouse studies in a futile search for "biomarkers" of aging, things like cholesterol levels and other blood chemicals that change with age—all to no avail. There was still no really good way to measure aging.

Ferrucci had been a geriatrician, a physician who treated older people, when he was recruited from his native Florence, Italy. In The Blast, Ferrucci saw a huge opportunity: There was no other study like it, no other that would allow him to plot the trajectory of individual subjects as they aged. He modernized the testing procedures and introduced new imaging technology like the MRI and CT scans, as well as broadening the study's scope.

In our first conversation, Ferrucci said something I would never forget: "Aging," he said, "is hiding in our bodies."

Aging is hiding in our bodies. What did he mean by that?

Two things. First, in terms of our biology, aging is a deep, almost subterranean process that begins long before we are even aware that it is happening; researchers believe some aspects of aging actually commence while we are still in the womb. These changes accelerate after we stop growing, at around age twenty, and by the time we realize that we are middle-aged (with or without the help of tombstone-shaped birthday candles) they are well under way. Many studies have shown that poor health at midlife is a direct predictor of both shorter lifespan and shorter healthspan. One study published in *JAMA* in 1999, based on Blast data, even found that simple handgrip strength in middle age predicted the onset of disability in old age. So Dr. Brown-Séquard had been right about that, too.

Second, aging is "hiding in our bodies" because we work hard to hide it, much as the previous owners of my hundred-year-old Pennsylvania house concealed its true rotten condition through the clever use of putty, shingles, and paint. According to Ferrucci, evolution has given us various ways to compensate for the effects of aging, "so these changes will have as few consequences as possible."

We compensate in mainly unconscious ways. For example, as the tumor grew inside Theo the dog, his heart compensated by working much harder, which is why he could still run around the block with me, just a week before he died. Or to pick a less awful example, studies have shown that people with a higher level of intellectual development—more education, more mental stimulation—are able to resist Alzheimer's disease for much longer than those of lower educational level. Their brains have developed more connections, and this stronger network is able to keep up the appearance of normal cognitive function, at least for a while.

But the most important domain of aging, Ferrucci believes, has to do with energy: how we store it, and how we use it. That was why Edgar had given me the mask treatment, that last morning. Our basal metabolic rate—how much energy one's body consumes while at rest, like a car that is idling—turns out to be an important measure of energy efficiency. The higher one's rate of "idling," the less energy is available for other needs, such as fighting infection or repairing tissue damage. I'd seen studies that linked higher basal rate with an increased risk of mortality (that is, death), so I was not pleased to recall that Edgar had murmured in my ear, "You have a high metabolism, don't you?"

Ferrucci was obsessed with energy efficiency, which is why so many of The Blast's tests measured physical performance, as if we were being scouted for the NFL—such as the quarter-mile hallway walk, for example, and the excruciating VO_2 max test on the treadmill. One of the most important tests, and one of the earliest indicators of aging, is the balance test—standing on one foot for thirty seconds and so forth. According to Ferrucci, that fine sense of balance is one of the first things to go. "I can go out and run ten miles right now," he told me (he's sixty and quite fit), "but I don't think I could do the balance test."

It may not seem like a big deal, but its ripple effect is felt literally for the rest of your life. As our balance weakens, we compensate by widening our stance, so to speak, placing our feet farther apart so as to provide a more stable platform. But this wider stance, in turn, makes walking and running much less efficient than our narrow, youthful stride. This, in part, is why older people seem to shuffle along, even when they are trying to run. As a result, we waste energy, and slow down even more. This is ultimately why walking speed—and walking efficiency—are so important, Ferrucci thinks. They are a sign of, basically, how much gas we have left in the tank.

There is a tragic irony here: We have less energy as we get older—and yet we use what we have much less efficiently. I couldn't help but think of my old dog Lizzy, who had by now passed her thirteenth birthday, which is pretty old for any kind of hound. She walked so slowly now that we sometimes had to step aside so that little old ladies with grocery carts could pass us on the narrow neighborhood sidewalks. She was running out of gas, right before my eyes.

Mobility is key to survival: This came up again and again in my research. Interestingly, this holds true all the way down to primitive animals, like the lowly *Caenorhabditis elegans*, a tiny, translucent, and sinuous little worm that lives in the soil. *C. elegans* is beloved by scientists because it contains most of the major body systems that we do, yet is smaller than a comma, so it doesn't cost much to feed. They're also convenient for studying aging because they only live for about three weeks. In the lab, their death is always preceded by a phase in which they simply stop moving.

Which explained why the quarter-mile walking test is considered one of the most important parts of The Blast: The slower you walk, the less energy you have in the tank (and, probably, the poorer your sense of balance). The real reason energy and mobility are so important, Ferrucci explained, is because they hint at other things taking place inside us that we can't see. Things happening even down at the cellular level. We slow down because we have more serious problems inside, like that rusty old Chevette on the freeway that can't seem to go faster than fifty miles an hour, even if its driver puts the pedal to the floor.

"You work harder, up to a certain point, but eventually you can't deal with it," Ferrucci said.

When we run out of gas completely, we reach a state called frailty, one of the end stages of aging. It doesn't refer to just fragility, but rather to a state of weakness and exhaustion, often characterized by slowness, low levels of activity, and unintentional weight loss. You waste away, basically, and at that point, it doesn't take much to push you over the edge.

Indeed, if my dog Lizzy reached the state of frailty, I would have to think about taking her on a one-way trip to the vet's office, just to be humane. Grandma and Grandpa aren't so lucky: For them, frailty marks the point where even a small problem, such as a routine illness or minor surgery, can quickly turn into a big problem, because the body simply can't bounce back from it. Because of his frailty, for example, my grandfather's urinary-tract infection set in motion a chain of events that ultimately led to his death.

Luckily for him, he wasn't frail for long. He could take care of himself until pretty close to the end, and handle most activities of daily life, even cutting his toenails—one of the most difficult tasks for older people, because it requires clear vision, precise movements, and, most of all, flexibility. It was a blessing, in a way, that he passed relatively quickly, without a long decline. But it was still horrible.

Meanwhile, his wife—my grandmother—is still alive as I type this, despite having barely done a single healthy thing in her life. Now aged ninety-seven, she lives in a nursing home in Florida and is nearly blind, yet still cooks and takes care of herself. Her breakfast is the same as it has been for decades: a sweet, delicious Danish pastry or, if she's really feeling adventurous, a bear claw doughnut.

I woke up the morning after my crab binge with a hop-tinged hangover and a mouth that tasted of Old Bay. It hadn't quite worked; I was still wallowing in middle-aged self-pity. I had to snap out of it: One of the more interesting Blast findings has to do with attitudes toward aging itself. Young-middle-aged people (in their forties and fifties) with positive feelings about growing older—gaining wisdom, freedom from working, opportunities to travel and learn more—tended to enjoy better health, and better cognitive health later in life.

Another significant body of research has found that overall, both men and women tend to grow *happier* with age. Some studies have actually pinpointed the mid-forties as the low point in terms of lifetime happiness—specifically, about age forty-six. It's like a U-shaped curve, higher in youth and in older age, lower in the middle. The conclusions basically supported what my mother has been telling me, that each decade of her life has been better than the last. Why couldn't I make myself believe her?

I began to regain hope after I met my next-door neighbor, an elegant African American woman in her mid-seventies whom I'll call Claudia. A retired federal employee from Washington, DC, which is also my hometown, she had been participating in the study for fifteen years, since she read about it in the *Washington Post*. Her presence alone was a mark of change: For its first twenty years, incredibly (or perhaps predictably), Dr. Nathan Shock's aging study was open only to white men.

But that was long ago, sort of; now the NIA actively recruited participants of both genders and all races, and Claudia was an old hand. We chitchatted about DC stuff for a while, and then she generously proceeded to give me the insider's lowdown on The Blast. I soon figured out her true objective in talking to me, however. She was dressed in a sporty-looking Ellesse tennis outfit, because she had just come from the treadmill test, designed to measure maximum heart rate and VO_2 max—how well the body takes in oxygen during intense exercise. It was the most feared test in the study, and also the most competitive. My turn was coming up, after lunch, and having done the test several times in the past I knew what kind of suffering it would entail. Yet Claudia hardly seemed to be sweating.

"Don't feel bad if I beat you on that test," she said, giving me a look. "It's pretty hard."

"It is?" I asked, my dread deepening.

She nodded gravely.

"You should know that they call me The Beast," she added.

Wait. Was I being trash-talked by a woman who was older than my mom?

Yes. Yes, I was.

But this was a good thing. In fact, it illustrated one of The Blast's key findings. Although it had been set up to discover uniform markers of aging, it ultimately found that, in effect, there are none. Aging is too varied, too chaotic, and too idiosyncratic—it's different for each individual, and the data reflects that. Initially, Blast scientists were dismayed by this fact; being scientists, they like their results to come out in nice, neat curves, with everyone basically falling in line.

Instead, they ended up with a plot that looked more like a shotgun blast, with data points all over the place:

WALKING SPEED

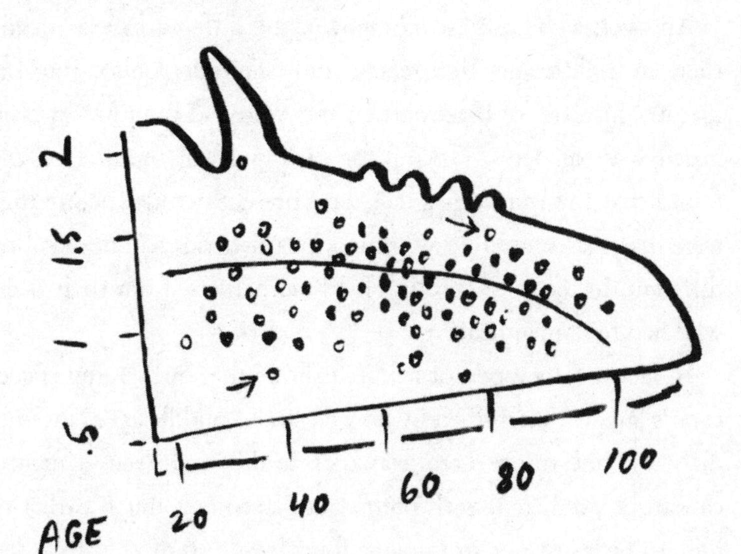

Of course, being scientists, they like to draw nice, neat curves on their charts, but Ferrucci is far more interested in the huge variations among individuals: Some eighty-year-olds, for instance, are barely mobile, but others seem to walk just as fast as the average forty-year-old. Indeed, some octogenarians actually walk faster than people three decades younger. Not that this actually happened, but it may even be possible that certain seventy-four-year-old African American women could kick the behinds of certain forty-six-year-old Caucasian males on the treadmill test.

Regardless, the point is this: We all age differently. Very differently. In fact, as Ferrucci observed, the paths we take in later life are far more divergent than the highly programmed process of our development. And the older people get, the greater the variation among them. Two random twenty-year-olds will have much more in common with each other, biologically, than any two seventy-five-year-olds.

And yet, as The Blast has shown, the differences are already there in middle age: In a recent study, Ferrucci looked into the medical histories of Blast participants who had been newly diagnosed with diabetes, looking for early signs of the disease. He found that the warning signals were present decades before they were diagnosed; even 30 years ago, the diabetics had been subtly different, in terms of certain blood biomarkers, from their peers who had remained healthy.

In fact, a large body of research shows that one's aging trajectory is largely determined by how we are in middle age. Not only diabetes, but future cardiovascular health and even dementia can all be predicted, with pretty good accuracy, much earlier in life. To pick just one example, a four-decade study that followed thousands of Japanese American men in Hawaii found that their late-life health was directly tied to certain key midlife risk factors. Those with lower blood pressure, LDL cholesterol, blood glucose, and body-mass index (BMI) in their forties and fifties, the study found, stood a much better chance of living to age eighty-five without any major health problems. By contrast, another large study found that obesity, high cholesterol, and high blood pressure vastly increased the risk of developing dementia later in life.

What's interesting is that all of these markers are pretty malleable; they depend largely on behavior and choices. Which tells scientists that much of aging is variable, and thus possibly changeable. "That's a wonderful thing; it's a window of opportunity," says Ferrucci. "If everyone was on the same deterministic biological trajectory, there would be no hope that we could change it. But the incredible variability shows that the potential to age well is there for everyone. And a few people are showing us the way."

I wanted to seek out those people, the seventy-year-olds who look, act, and test out like they are fifty or younger; those were the ones I wanted to draft for my senior-citizens-league softball team. And then I realized that I already knew at least one such individual extremely well: my own father.

Chapter 5

HOW TO LIVE TO 108 WITHOUT REALLY TRYING

If you live to be one hundred, you've got it made. Very few people die past that age.

—George Burns

I was standing on a street corner in Lower Manhattan one day a few years ago, waiting to meet my father for lunch, when I spotted what looked like a scarecrow walking toward me. His dark suit hung off his shoulders and flapped in the breeze as he walked, but even from a block away I recognized my dad's distinctive swinging gait.

We look a lot alike, my dad and I. We share the same thinning blond hair, same khaki-based white-guy style, even the same name. As he plumped up in middle age, I would inherit the pants and jackets he'd gotten too stocky to wear. Decades of worldwide business travel and late-night Manhattan restaurant dinners—literally, 250 to 300 a year—will do that to a guy. The man walking toward me now could have been a completely different person. And in some ways, he was.

When he reached his mid-sixties, my father made a series of radical life changes that he seemed to hope would help him beat some of the effects of aging. He had reason to worry: His own brother had died suddenly of cancer before he turned fifty, and his parents had not fared very well, either. After sixty-nine years of eating a steak-based Midwestern diet, and a lifetime of chain-smoking, his father had found out that he needed triple-bypass surgery—one of the surgical "fixes" that have helped push life expectancy beyond the traditional first heart attack in one's fifties. He survived the operation, and my grandmother cared for him for two years. Then one day, as she got ready to take him to a doctor's appointment, she lay down for a quick nap and died of a massive, unexpected heart attack, at just seventy-one.

We haven't really talked numbers, but my impression is that my dad would rather check out closer to age one hundred—killed instantly by a flying golf ball, perhaps. He'd already had one scary brush with illness, not long before our lunch, and while he seemed to be out of the woods, he not surprisingly hates the very idea of getting old, of sickness and death. There will be no going gently into that good night. His end-of-life care plan, in case he ever ends up incapacitated or demented in a nursing home, unable to recall who won the 1963 U.S. Open, consists of three words: "Just shoot me."

My siblings and I are pretty sure he is serious.*

So at age sixty-seven, he had fled New York City and its stresses and moved back to his native Illinois, where he essentially remade his life. He now spends his days pursuing his great passion, which

* My brother the corporate lawyer has been appointed to do the deed, if it ever comes to that.

is golf. An accomplished chef and former devoted carnivore, he has switched to a largely vegetarian diet, à la Bill Clinton, hoping to avoid the calamitous death rate associated with meaty, fatty Western fare. Unlike Clinton, he'd never liked Big Macs, but I was shocked to learn that my father had also largely given up wine; after all, this was a man whose idea of the perfect father-son trip was to drive around Burgundy, feasting at Michelin-starred restaurants. (I was sixteen at the time.)

I was lucky; I grew up with the prototypical hipster dad. Embarrassing as it is to admit, he introduced me to some of my lifelong favorite bands when I was a teenager and he was in his forties. He was a devoted early adopter, always bringing home the latest technological gadgets, such as the huge, brick-like Motorola cell phone that Michael Douglas brandished in the ur-'80s movie *Wall Street*. A lifelong information junkie, he researched his various interests obsessively, and over time he had amassed a formidable library, including hundreds of books on food and cooking, and more than 1,000 volumes on golf. It rubbed off on me: Instead of throwing me a football, he took me to the public library, where I buried myself in books and developed the strange urge to one day write them.

Now he had become obsessed with his health, seeking the latest diets and anti-aging strategies to extend his lifespan and, more important, his healthspan. Four times a week or so, he hops on his new bike and rides over to the local rail-trail, where he'll hammer across the prairie for an hour and a half. On the other three days, he might go to his golf club to play a fast-paced round (walking, never in a cart). Or, if the weather is bad, as it often is in Chicago, he will abuse his rowing machine for a solid hour while watching DVDs of college lectures. For dinner, he and his age-appropriate lady friend will eat something veggie-based, or perhaps split a six-ounce piece of fish plus a salad. As for snacks, he eats precisely twenty-three almonds every day.

At lunch, he asked if I wanted any of the pants that he was now too skinny to wear. I thought I detected a smirk. Instead of legal papers, his briefcase was loaded with supplements, including fish oil, coenzyme Q10 (thought to be good for heart function), as well as resveratrol, the powerful compound found in red wine that has been shown to extend lifespan in obese mice. He subscribed to an array of health-related magazines and newsletters, including *Life Extension* magazine, which ran long feature articles touting various supplements (and is published by a company that sells those same supplements).

His latest find was curcumin, a derivative of the spice turmeric that is a staple of traditional Ayurvedic medicine. More recently, curcumin has racked up some fascinating results in the lab, suggesting that it might work against a wide range of ailments, from diabetes to irritable bowel syndrome to some kinds of cancers, including colorectal cancer. There have been encouraging preliminary studies, particularly against inflammatory conditions; and in the lab dish at least, it also seems to zap cancer cells. But the evidence is far from conclusive, and there have been no large-scale randomized clinical trials in humans; the jury is still very much out. Nevertheless, better safe than sorry: Dad takes eight *grams* of it each day, or about half a tablespoonful, which seems like a lot.

He certainly seemed healthier than the last time I'd seen him; his skin sort of glowed, although the only discernible effect of all the curcumin was that it turned certain bodily functions bright yellow. (I was sorry I asked.) Was it also beating down any incipient cancer genes that might be lurking in his cells? Who knew.

Still, who could blame him for fearing decrepitude, disease, and a long-suffering death? If his somewhat odd new habits help stave off the dreaded "just shoot me" day, then it will be well worth it. Certainly, they can't hurt, although Jay Olshansky would point out that he also benefits from the two most powerful anti-aging drugs known to man: money and education.

So far, he seems to be headed in the right direction: Even now, in his early seventies, he can drive a golf ball so far and so straight that guys half his age stop in midswing and go, *whoa!* He rode more miles on his new $3,000 bike in six months than I did the entire year, and I used to race competitively. More importantly, he had been able to throw away his blood-pressure pills, and now he takes no medications whatsoever. He is so healthy, in fact, that he too has been accepted into the Blast study, an accomplishment that made him no end of proud (and which reduced me to the low level of lying about my results in certain tests). All in all, he seems well on his way to sticking around long enough to spend every last penny of his grandchildren's inheritance.

But as his concerned and congenitally skeptical journalist son, I couldn't help but wonder: *Will any of this stuff actually get him to a hundred?*

Not according to Nir Barzilai. For more than a decade, Barzilai has been studying a cluster of Ashkenazi Jewish centenarians who live in the New York area, which has led him to some remarkable conclusions. An affable jokester with thick glasses and a babyish face, the fifty-six-year-old Barzilai could be an older brother to Mike Myers's Austin Powers character, but beneath his schmoozy manner lurks the penetrating mind of a scientist on a mission.

He trained as an endocrinologist in his native Israel, and as a young physician in the 1980s he worked in clinics in Soweto, the poor black townships outside Johannesburg, South Africa. Even now, he still makes time to see patients, to keep himself grounded in the problems of real people. As his interest shifted to aging, he began wondering why most people become sick when they get older, while a select few almost seemed to be immune from aging-related diseases. He decided to focus on the longest-lived among us, the centenarians. What makes them different?

Recruiting subjects was not easy, he found; elderly Jews tended to have complicated feelings about medical research. As the only child of parents who had survived the Holocaust, Barzilai was sensitive to their concerns. After spending countless Shabbats visiting synagogues all over the New York metropolitan area, he recruited more than five hundred elderly Jewish folk—they must be ninety-five or older to qualify—into his unique study of aging. (He also jokes that he will accept healthy hundred-year-old converts.) We'll call them the Super*Bubbes*, using the Yiddish word for "grandmother," since most of them are women anyway. Every year or so, he and his lab assistants subject the Super*Bubbes* to an array of physical examinations, cognitive tests, lifestyle questionnaires, and blood analyses, rather like an All-Star version of The Blast.

As a rule, he found, the Super*Bubbes* do not generally eat "healthy" diets and compete in triathlons, or engage in any other kind of recognizable exercise, for that matter. When they're hungry, they'll reach for a knish, not quinoa (sorry, Dad). Not only that, Barzilai has found that many of his Super*Bubbes* have actually smoked, some for decades, like Madame Calment. Nearly half were overweight or even obese, while less than 3 percent were vegetarian. Nevertheless, they ace their blood tests; in particular, they tend to have spectacularly high levels of HDL, or "good" cholesterol. "They have the *best* blood you have ever seen," Barzilai says, excitedly. "Their blood is perfect!"

It's not that Ashkenazi Jews have any special advantage in terms of longevity; on the whole, they don't necessarily live longer than any other group of New Yorkers. Barzilai studies them because, for one thing, the strong cultural identity and long history of intermarriage among the *Ashkenazim* means that their genomes are relatively similar to one another. And that, he thinks, will help him pick out the "longevity genes" that he thinks differentiate the centenarians from the rest of us.

Barzilai's theory was that the centenarians live longer for the simple reason that they age more slowly. That may sound obvious, but the interesting question is *why*. If you could somehow tease out which genes are actually responsible for their slower aging, then you'd be on to something major. "Most biology is about ways in which we are the same," he told me. "This is an unbelievable opportunity for us to understand why the biology of some is different from others."

Biologically speaking, men don't get much more unique than Irving Kahn, whom I met one sunny November morning in his corner office, twenty-two stories above Madison Avenue. He was just finishing up with the *Wall Street Journal* and the *Financial Times*, and preparing for another day as head of his family's investment firm, which manages some $700 million in assets. This was impressive, because Kahn looked to be well into his 80s, maybe 90—but in fact, he was 106 years old when we met. In three weeks, he would turn 107. "Thirty-seven people are coming to my birthday party," he rasped. "I have no idea why."

Perhaps they were coming for the same reason I was there: Because Kahn is one of the oldest men alive. He was born in 1905, three years before Henry Ford produced his first Model T. His family lived in Yorkville, on the Upper East Side of Manhattan, when it was largely populated by Polish and Hungarian immigrants. His father worked as a salesman, selling fine chandeliers to wealthy New Yorkers; oftentimes, he had to first persuade them to install electricity in their homes. Irving went to work on Wall Street in 1928, and ended up as a disciple of the legendary Benjamin Graham, the godfather of analytical investing. He lived through the Crash of 1929 in his first year on the job, and he considers Warren Buffett, whom he met and mentored after World War II, to be a mere pup. He is a survivor, a human version of The Ark, my family's everlasting cottage on the lake.

To call Irving Kahn an extreme outlier is understating things. According to Barzilai, only one person in ten thousand makes it to the century mark, and three-quarters of those who do are women—despite the fact that very old women, in general, tend to have more health problems than very old men. "The women are less healthy," he observes, "but they live longer."

Indeed, this appears to be true at all stages of later life: One large cohort study of eighty-five-year-olds in Newcastle, England, found that men actually reached that age in markedly better functional health than women. One in three of the eighty-five-year-old men could perform every one of a list of seventeen different tasks of daily living, without help (things like brushing teeth and bathing), versus just one in six of the women. But the men still died sooner. "The men drop dead," says Thomas Kirkwood, director of the Newcastle study, "while the women keep going."

But very few women or men keep going much past a hundred,

and for those who do, the risk of dying each year begins at one in three, and goes up from there. According to the Social Security Administration's life expectancy calculator, as of our December 2012 meeting, Kahn had just 18 months left to live, statistically speaking. But who cares about statistics? Kahn has already defied massive odds to become one of the oldest living Americans, pushing right up against the observed limits of male longevity, and yet he still goes to work each day. "There are how many people like him in the world?" asks Barzilai. "I don't know, ten? Twenty, maybe?"

Not only that, but Kahn's three siblings *also* lived past 100, including his older sister Happy, who passed away in 2011 at the age of nearly 110, despite having smoked for 95 of those years, à la Madame Calment. Irving himself had smoked for about thirty years, with no observable ill effects. Yet he seemed to regard his extreme age as a sort of mild annoyance, rather than the near-miracle that it is. He missed being able to walk the twenty blocks or so to work every day from his apartment on the Upper East Side, as he used to do until a building fire drill in 2002 forced him to descend twenty-two flights of stairs. His knees have bothered him ever since. He's even more ticked off that, as he put it, "my eyesight is beginning to diminish." How's he supposed to read the *Journal* every day?

The stereotype of old people being kept alive by doctors and drugs, as they endure a living hell of medically prolonged infirmity, just doesn't apply to centenarians like Irving Kahn, Barzilai insists. Oftentimes, he notes, when they come in to be examined for his study, "it's their first time going to the doctor." They've just never needed to go, but it may also help contribute to their continued survival, Barzilai thinks. "The oldest person in the world has steadily been getting younger," he told me, "and I think it's because the doctors have been killing them," with unnecessary treatments like statin drugs for cholesterol, when in fact they are extremely unlikely to suffer a heart attack. So my Christian Science relatives might have had a point, after all.

Because centenarians stay healthier for longer—another hallmark of slower aging—they are also a lot cheaper to keep around. Someone who dies at a hundred will rack up just one-third the medical bills, in their last two years of life, as a person who dies at age seventy, according to the Centers for Disease Control. Moreover, the seventy-year-old will have been sicker for much longer, an average of seven years of ill health at the end of life, versus less than two years for a centenarian. Public-health types call this the "compression of morbidity," shortening the period of time during which old people are sick. This is a good thing: More lifespan *and* more healthspan, the opposite of the Struldbrugs. And, obviously, Irving Kahn has it pretty much figured out. But how?

He wasn't about to divulge the secret, whether he knew it or not. His late-life fame and the attendant publicity—he's been quizzed by TV journalists from all over the world—had hardened him into a wily, evasive interviewee. I was getting nothing. "Irving, am I correct in saying that you find your age is not particularly remarkable?" his grandson Andrew prompted him, in a louder-than-normal voice to compensate for his grandfather's somewhat diminished hearing.

Irv nodded.

"And Bill, you might ask a question," Andrew continued, "and Irv, you will do your best to evade the answer." His tone verged on accusatory. "You will talk about a wide range of topics that have nothing to do with the question!"

Which is what Irv proceeded to do for the next two hours,

leading me on a wild ramble from subject to subject. His interests are wide ranging: He remains a voracious reader, but only of nonfiction. Novels and poetry he regards as a waste of time, though he makes an exception for Shakespeare. My attempt to interest him in my ninety-seven-year-old grandmother, a looker who still puts on her pearls and makeup every day, fell flat; he's still enamored of his wife, Ruth, who died in 1996.

We talked about the stock market, inevitably, and his never-ending search for undervalued businesses. He bought one company, a shipping conglomerate called Seaboard, for a few dollars a share, decades ago; now it's at $2,600. Buy-and-hold makes a lot of sense as an investment strategy when you live to be 109* (as of December 2014). And it fits with one of Barzilai's other findings, that his centenarians tend to have positive attitudes about life in general, in addition to their spectacular cholesterol numbers. As Irv put it, "The alternative is not useful."

The fact that he goes to work every day, incidentally, is both a function of his extreme longevity—and also, in part, perhaps a contributing factor. On Okinawa, the famed Japanese "Blue Zone," with the highest concentration of centenarians in the world, older people talk about the importance of *ikigai*, which translates as "a reason to get up in the morning"; in short, it's your purpose. Irv's *ikigai* is to find the next Seaboard. I was merely keeping him from his work.

He told stories about Ben Graham, and about his childhood experiments with the newfangled technology of radio. "Is that the Victrola?" his mother had asked. And so on like this, for two hours. He'd witnessed inconceivable changes during his

*Irving Kahn died in February 2015, one week after the publication of this book.

incredibly long life, yet his own longevity was the one thing that he, of all people, could not explain. "You're asking questions that have no good answer," he growled at last. "I don't think I can tell you what you want to know."

"Irving is our poster child," Barzilai told me, a few days later, "but biologically he is not interesting to me anymore, because he is at the end of his life. You really want to know—when he was fifty, sixty, seventy—how he was."

The best he could do was to study the offspring of Irv and his fellow centenarians; he had found they, too, seemed to be aging more slowly than their peers. So Irving's son Tommy, who came in to greet us, looked more like fifty-one than his actual age of seventy-one; and his grandson Andrew, who was in his thirties, would not have looked out of place in a Boy Scout uniform. And of course Irv's three siblings had all made it past one hundred as well. The reason they all seemed to age slowly had to lie in their genes, Barzilai felt. I silently hoped I had received some DNA from my own long-lived ancestor, the ninety-six-year-old Puritan Elizabeth Pabodie.

The role of genes in determining longevity has been debated as long as we've known that genes exist. Studies of Danish twins have revealed that longevity is only 20 percent inherited, and 80 percent due to environmental factors. But that holds true only up until about age eighty-five. After that, heredity moves to the forefront, making up more like half the puzzle, if not more. As the evolutionary biologist Steven Austad (whom we'll meet in chapter 7) puts it: "If you want to become a healthy eighty-year-old, you need to live a healthy lifestyle, but if you want to become a healthy hundred-year-old, you need to inherit the right genes."

The fact that all four Kahn siblings lived so long is utterly unsurprising to Barzilai. "We are certain, there is no doubt in our minds, that exceptional longevity is mainly inherited," he says. Just not in the way that he first suspected.

Barzilai initially believed that his Super*Bubbes* possessed "perfect" genomes, their genes optimally tuned for longevity. But as gene-sequencing technology developed, along with our understanding of the role of genes in disease, he was surprised to discover that the opposite was true: Many of his centenarians actually possessed some of the same crappy genes as the rest of us. He and his team sequenced the genomes of forty-four centenarians and found that nearly all of them possessed undesirable gene variants that are thought to promote nasty conditions, including heart disease, Alzheimer's, and Parkinson's. Yet none of them had gone on to develop any of those diseases. This led him to ask a different question: "How come people who should have dementia at seventy, and be dead at eighty, are living to a hundred?"

Instead of "perfect" genes, Barzilai decided, the long-lived must instead have *protective* genes, which prevent them from developing the usual diseases of old age. (Such ultraprotective genes may also explain why Keith Richards is still alive.) He decided to zero in on their cholesterol-related genes, because his centenarians had such excellent blood values and perfect cardiac health. He found that many of them had a specific gene variant that inhibits something called CETP, a molecule involved in cholesterol processing (it stands for "cholesterol ester transfer protein," if you want to geek out on it). It's complicated, but in general, the less CETP you have, the better; high levels of the protein are thought to lead to premature atherosclerosis. Centenarians with the CETP-inhibiting mutation not only had better heart health, and very high "good" cholesterol, but they also had a lower incidence of memory loss and dementia, according to a paper Barzilai and his colleagues published in *JAMA*.

This single gene, in other words, seemed to be protecting them from two of the Four Horsemen, heart disease and Alzheimer's. Irving Kahn has the CETP variant, as did his siblings—along with roughly one in four of their fellow centenarians. But among sixty-five-year-olds, only one in twelve has the gene, which means that those with the gene are three times more likely to live to be a hundred than those without.

This finding dovetailed with research being done by the pharmaceutical companies, which were already desperate for a drug to replace the statin class of cholesterol-lowering medications, whose patents were all expiring. Merck and Pfizer and others raced to develop CETP inhibitors, which were meant to raise HDL ("good") cholesterol. But so far, those drugs have not yet panned out, and some studies were stopped because too many patients died. (Merck has one still in Phase III trials, but it is the last survivor in the class.)

"They made a dirty drug," says Barzilai, one that affected too many other targets besides CETP. He still believes that by finding more protective genes, and developing treatments that mimic their effects, it might be possible to extend healthy life for the rest of us—and "bypass the complexity of aging," as he put it. The search is only beginning.

One other possible longevity gene had to do with IGF-1, the growth factor—which may not be so good for you, as we've seen in chapter 3. Barzilai found that his centenarians had relatively high levels of IGF-1, but oddly, their cells were actually resistant to it—also because of another gene variant. That is, Irv and his fellow oldsters aren't just small because they are a century old; they have lived so long *because* they are on the smaller side, sort of like human Chihuahuas.

Yet the most striking thing about the longevity genes discovered to date is how rare they are, even in centenarians. Barzilai and other research teams have only been able to identify a handful of candidates, which he blames on the high cost of genomic-sequencing technology—and also on the relatively tiny number of centenarians worldwide. There are so few of them, relative to the rest of us, that it is difficult to single out specific genes as a statistically-convincing "cause" of their longevity. That could change, as sequencing becomes cheaper over the next decades. Barzilai and his team are working with GoogleX on a gene-sequencing venture, and in March 2014 geneticist Craig Venter jumped into the game, forming a company called Human Longevity, Inc., which promises to sequence forty thousand genomes in a hunt for clues to the genetics of aging.

But for now, I had a different question: If there really are such things as longevity genes, why don't we *all* have them? Why hadn't evolution blessed all of us with these wondrous, cholesterol-improving, brain-protecting, slow-aging, life-extending genes?

But the deeper I looked, the more I realized that, in fact, most of us have the opposite of longevity genes—and that much of our DNA dearly wants to kill us.

Chapter 6

THE HEART OF THE PROBLEM

Men are born soft and supple; dead, they are stiff and hard. Plants are born tender and pliant; dead, they are brittle and dry. Thus whoever is stiff and inflexible is a disciple of death. Whoever is soft and yielding is a disciple of life. The hard and stiff will be broken. The soft and supple will prevail.

—Lao Tzu

"You're gonna get stabbed a number of times," Bill Vaughan said by way of greeting when I appeared at the door of his house in the Berkeley Hills. "But it's all in the interest of research."

Despite only having known Vaughan for twenty-four hours, I was already accustomed to taking abuse from him. When we'd sat down to lunch the previous day, he'd announced, "You want to understand aging? I'll show you aging!"

And with that, he'd pinched the back of my hand, hard. It felt like a hornet sting. As I looked at him in shock, he did the same to himself.

"Look!" he said, thrusting his hand almost in my face. "See the difference?"

On my hand, the pinch mark had disappeared almost instantly, like a ripple on a pond. But on his hand a persistent little fold of skin lingered.

He leaned back in his seat. "That's aging!" he said. "It's *elasticity*—the loss of elasticity. That's what it's all about."

The difference came down to two molecules: collagen, a tough, rubbery substance that gives structure to lips and tendons and skin; and elastin, which lets our skin snap back from a smile or a frown, rather than remaining there as a wrinkle. But Vaughan wasn't talking about wrinkles. Elasticity, to him, is a much broader concept, maybe even the key to aging. "It affects lung function," he said, "it affects heart function. And we still don't really understand it."

Vaughan was seventy-seven but didn't look it, with his shock of dark hair and broad, unlined face. My first impression was that he sort of resembled the actor Christopher Walken, which turned out to be some of his friends' first impressions of him, too. In the early 1980s, as an underemployed biochemistry PhD, Vaughan had used his wife's kitchen mixer to whip up a kind of gluey energy food for athletes that eventually became known as the PowerBar. He sold out to his business partner and co-inventor before PowerBar became a household name, but he never regretted it. "He just worked all the time," he says of the partner, whose name was Brian Maxwell. "And he dropped dead of a heart attack when he was fifty-one."

Pause.

"In a post office."

A few years later, Vaughan revved up the KitchenAid again, this time looking for a better way to feed his daughter Laura while she was running (and winning) ultramarathon races of one hundred miles or more. She needed to get calories into her body, but her stomach could not handle any kind of solid food while she was running, often for twenty-four hours at a time. Not even a PowerBar ("It just sits in your stomach," Vaughan said). So he came up with a kind of sugary gel, enhanced with a secret mix of vitamins, electrolytes, amino acids, and herbs, that he would dispense to her from plastic flasks at key points along the course. Vaughan eventually dubbed the mixture GU, and it hit the market in 1994 as the world's first athletic energy gel. Sold in little foil packets to runners, cyclists, and any other athlete in need of a quick shot of fuel, GU changed the world of sports nutrition as radically as the PowerBar had.

Today there are dozens of energy bars and gels on the market, but both billion-dollar product categories originated in Vaughan's own kitchen. "I'm a problem solver," he told me. "The problem that I had to address was that my kids were going off to running practice at six in the morning, with nothing in their stomach, and then not eating until noon. They were not created with marketing in mind."

He eventually sold GU to his son Brian, which let him retire to his house in the Berkeley Hills and turn his attention to "the ultimate problem," the one that had consumed his mind since he was a graduate student: aging. His homegrown basement laboratory was famous among Berkeley's elite athletes, who often come there to be tested for their performance potential—but it is even better known among the Bay Area's large and growing community of amateur aging scientists and do-it-yourself "health hackers," who monitor their own vital signs and take action accordingly. To the "quantified self" movement, the folks who track themselves with Fitbits and complicated scales and such—and even send out their own feces to be analyzed—Vaughan is like patient zero, because there in his basement lab he has been doing his own self-studies for decades. I was dying to see it.

He was serious about the research. The shelves of his downstairs office were lined with notebooks, some marked ATHLETES, others AGING STUDIES; a printout of his wife's latest test results lay atop another binder labeled FAMILY. On a lab bench in the next room sat the white KitchenAid mixer that had given birth to both the PowerBar and GU. Vaughan had gotten into self-testing while he was still a graduate student, way before anyone had ever uttered the phrase *quantified self.* He was studying biochemistry in the building where HDL cholesterol was first identified, and like his fellow graduate students he was often drafted as a guinea pig for studies. He caught the research bug, and the notebooks on his shelves now held five decades' worth of data on his own blood chemistry—his own, personal version of The Blast.

He isn't selfish, either. Once in a while, he hosts parties where his aging-conscious friends get their blood drawn and compete at the "lung function test," a surprisingly painful exercise where you take in as much air as possible and then blow it out—all of it—into a white plastic tube connected to a computer. I'd done the same thing at The Blast, and it was about as much fun as trying to make yourself vomit. The test is supposed to measure lung capacity—which depends, of course, on elasticity. (The older we get, the stiffer our lungs.) Those results go into the notebooks, too, and I silently prayed that Vaughan served a lot of good booze at these "parties."

We got down to business quickly. First up was the cholesterol test. Within minutes, Vaughan was swabbing my middle finger with alcohol and then—*ka-chinch!*—punching a hole in my fingertip with a disposable plastic lancet. "Let's see if you're a bleeder," he said, holding my finger firmly as a drop of blood began to form. "Okay, good."

In a smooth, practiced motion, he suctioned off the blood into a thin glass pipette and transferred it to a small rectangular cartridge, which he then placed in the open tray of a small beige machine that looked like a parking-ticket validator but was, in fact, a state-of-the-art blood analyzer, normally found only in high-end medical offices. The machine began whirring.

Vaughan performed this ritual on himself at least once a week, he said, and the benefits more than justified the machine's $1,300 price tag. Rather than having to go to the doctor's office to get blood work done once a year, he simply does it himself, whenever he wants. "I use it to see whether my various dietary interventions have any effect," he explained. "A DIYer like me can find out what works and what doesn't in a few weeks—what works for *me*, not as a statistic."

Like Suzanne Somers, he's become his own personal science project, the main difference being that Vaughan has an actual science degree (also, he skips the hormones). He pays special attention to his LDL cholesterol—the "bad" kind—which tends to be extremely low in centenarians. If he keeps his LDL low, he reasons, then he might stand a chance of living to be a hundred, too. The machine lets him monitor it closely: If it gets too high, he'll cut back on the carbohydrates and perhaps up his dose of red yeast rice, a natural statin that lowers LDL. He used to run a lot, until his aging knees forced him to quit, and now he spends a good chunk of each day sitting on a Lifecycle, scrolling through the latest research studies on the PubMed database. When his cholesterol drops down to a comfortably centenarian-like level, he half joked, "I eat a piece of cheesecake."

"Maintenance is tough," he admitted. "You gotta be on it, as you get up there, and you've got to spend much more time on it." He spends nearly all his time either combating aging, via exercise and yoga, or studying it: I'd run into him at conference after conference, and he was always up on the latest research.

The machine beeped, and the digital readout flashed to life. "There you are!" he said, leaning in to check the number.

"Two hundred thirty-five," he read slowly. "Wow!"

This was not a good "Wow!"

According to the latest guidelines from the American Heart Association, a total cholesterol level of more than 240 is defined as dangerously high; ideally, it should be less than 200. My numbers got worse from there. My HDL cholesterol—the "good" kind, the higher the better—came in at fifty-six, which was okay but not great. Next was LDL, the one I'd been waiting for. Under one hundred is good. Vaughan has beaten his down to around thirty-five, which is centenarian territory. Mine rang the bell at a gelatinous 154.

Maybe I shouldn't have eaten that Double-Double at In-N-Out Burger last night, I thought to myself.

"That's high," he said, frowning.

"That *sucks*!" I yelped.

"Yeah."

Three months after my visit to Bill Vaughan, I found myself sitting in a waiting room in Fort Lee, New Jersey, crowded shoulder-to-shoulder with more than a dozen senior citizens who were kvetching in at least four different languages. It was close to ten on a Friday morning, and we were all here to get our blood tested. Things had not been going smoothly, and the crowd was getting restive. "She got here *aftah* we did!" one well-made-up matron snarled as another lady was ushered into the back room. Meanwhile, the middle-aged woman next to me reassured her septuagenarian mother, "You *want* your doctor to be younger than you are."

I had come to see a youngish cardiologist named Nathan Lebowitz, to try to get a handle on the whole cholesterol issue. High cholesterol seems to run in my family—my grandfather's had hovered up around three hundred, and my little visit to Bill Vaughan's home lab had finally motivated me to get checked out. I was eager not to join the roughly six hundred thousand Americans who die of heart disease every year, more than from any other cause. Cancer runs a close second, but here's the thing: Every one of those cancer patients, as well as stroke victims, and respiratory-disease sufferers, and the rest, also died because, eventually, their hearts stopped beating. Your heart quits, and you're done. Obviously.

I picked Lebowitz not because he regularly made "Top Docs" lists but because he focused strongly on prevention, which studies had long identified as key to fighting heart disease. He was also known for taking his time to explain things to patients, far more than the six minutes of face time that insurance companies typically allow. I didn't need to hear my doctor tell me that my cholesterol "should be lower" one more time. I wanted to know more. "You can catch things incredibly early," Lebowitz had said on the phone, still sounding like the native Brooklynite he is. "Why wait?"

Why indeed. Everything we know tells us that heart disease has a long, long prologue. In one famous study from the 1950s, serious coronary arteriosclerosis—thickening and stiffening of the main artery leading from the heart—was found in 77 percent of a group of three hundred patients. These were not old or even middle-aged men, either: They were young soldiers who had been killed in Korea, and their average age was just twenty-two. Yet serious heart problems often go undiagnosed; in two-thirds of male patients, the first sign of heart disease is actually a heart attack. For women, that number is more like half, which is still alarmingly high.

Much of what we know about heart disease comes from the famous Framingham study, which is like a heart-focused version of The Blast that has been going on for more than fifty years in Framingham, Massachusetts, covering multiple generations of people. It was Framingham that first revealed the link between high cholesterol and the likelihood of heart attacks, along with four other key risk factors: obesity, diabetes, high blood pressure, and smoking. Scientists have been poring over Framingham data for decades, but in 2006 a team of epidemiologists laid out the bottom line in a major study published in *Circulation*, the journal of the American Heart Association: "Even the presence of a single major risk factor at 50 years of age is associated with substantially increased lifetime risk of CVD [cardiovascular disease] and markedly shorter survival."

Gulp.

So as I drew near the magic tipping point of fifty, I found myself the proud owner of not just one, but two out of the five Framingham risk factors: high-ish cholesterol and not-so-low-ish blood pressure (140 over 90, and climbing with each successive draft of this book). Even though I fancied myself some sort of athlete, I was still not immune. Legions of middle-aged athletes have keeled over dead from heart attacks, including the running guru Jim Fixx and thousands of others whom you've never read about.

My sky-high LDL number was particularly worrisome. The problem is that goopy LDL particles tend to get stuck in tiny fissures in the artery walls, creating "plaques" that then capture more cholesterol and other cellular junk as it passes by in the bloodstream, the way a tree falling into a river will snag all sorts of debris in its branches. Those plaques can then harden and constrict blood flow, a condition called atherosclerosis (which is different from arteriosclerosis, which is hardening of the artery tissue itself). The plaques are full of bad stuff, and if one of them should suddenly rupture and break off, the bad stuff travels directly to the heart, and then it's good-bye, Uncle Billy.

For the last few decades, the main goal of cardiology (and pharmacology) was to try to get those cholesterol numbers down, usually via statin drugs, which cut down LDL; our frenzy to lower cholesterol quickly made statin drugs like Lipitor and Crestor the best-selling class of drugs ever. And we succeeded, to a remarkable degree: Since 1960, death rates from cardiovascular disease have been cut in half.

But there had to be more to it. For one thing, while statins did appear to help patients with existing heart disease, who were already sick, widespread use of the drugs for prevention has not really lowered overall mortality rates among healthy people. That is, statins only seem to help those with existing heart problems, but don't work as primary prevention. One explanation could be that statins appear to increase one's risk of diabetes—by as much as 26 percent, according to one recent large study. "You should see a decline in overall mortality, and you don't," says Nir Barzilai. "So that, to me, means that statins kill them in another way."

More recently, there's been some doubt as to whether cholesterol is really the whole story. A major study of 136,000 patients who had experienced a coronary "event" found that half of them actually had *low* LDL cholesterol, according to then-current guidelines. TV newsman Tim Russert was one of them: When he died of sudden heart attack in 2008, his LDL was a saintly sixty-eight. Although his weight was an obvious risk factor, and he was on blood pressure medication, he betrayed no serious symptoms before an arterial plaque ruptured and killed him at age fifty-eight.

This is not to say that heart disease has no warning signs. High blood pressure, which Russert had, is an easy one. Less obvious is simply whether or not someone *looks* old. Data from the decades-long Copenhagen City Heart Study shows that people who display certain outward signs of aging, such as fatty deposits around the eyes, creased earlobes, and baldness or receding hairline (uh-oh), had more than a 50 percent greater risk of a heart attack. Another one: Slower reaction time, which Bill Vaughan also measures in his basement lab, has also been shown to predict risk of death from cardiovascular disease. And finally, there is one bellwether of heart trouble that on reflection should be obvious: erectile dysfunction. Viagra was originally developed as a blood-pressure medication, before its other life-enhancing effects were detected. A 2012 study showed that, in fact, men with more severe erectile dysfunction were also more likely to have preventable heart disease. Like the ads say, "It could be a question of blood flow."

But cholesterol remains the most obvious, most quantifiable risk factor. The bad news is that, in all likelihood, your doctor (like mine) has been measuring it the wrong way. "The regular cholesterol blood test that we've been doing forever, that's not the real story," Dr. Lebowitz said when I finally managed to get in to see him. "It's an attempt to represent the real story."

He began by setting me straight on one big point: Not all cholesterol is bad. Cholesterol is actually a molecule essential to life, crucial for making cell membranes as well as producing hormones like testosterone and estrogen; that's why it's in our blood to begin with. Cholesterols are also required for brain function, so it's neither possible nor desirable to Lipitor them completely out of existence. Moreover, cholesterol comes in all sorts of shapes and sizes, not just the big three (good, bad, and triglycerides). And just to make things even more complicated, he said, not all bad cholesterol is so bad—and not all good cholesterol is good.

Lebowitz opened a folder and peered at my lab results. To get a better idea of his patients' true risk profile, he relies on a sophisticated array of blood tests called the Boston Heart panel, which measures a bewildering array of parameters. My results took up three full pages, but they told a slightly more hopeful story than Bill Vaughan's home blood-test machine. Since my visit to Berkeley, my LDL had dropped; that was the good news. The not-so-good news is that it had only fallen by 6 points, from 154 to, erm, 148.

Lebowitz frowned. Cholesterol numbers can bounce around from day to day, but this was pretty bad. Still, all hope was not lost. The important thing with LDL is not just the total number, he explained. All that LDL has to be carted around in special carrier molecules, which vary considerably in size. As he put it, a small number of large carrier molecules is better than a large number of small carriers, in the same way that it would be safer to transport fifty tourists around Rome in a single large bus rather than on fifty separate mopeds. Just as the mopeds would be much more likely to get into accidents, a bunch of small carrier molecules would have more opportunities to stick into my artery walls and cause plaques that might eventually kill me. "Just playing the percentage odds, more of them are gonna hit your artery wall, and some of them are gonna stick, and wiggle their way into the cracks, and cause problems," he said.

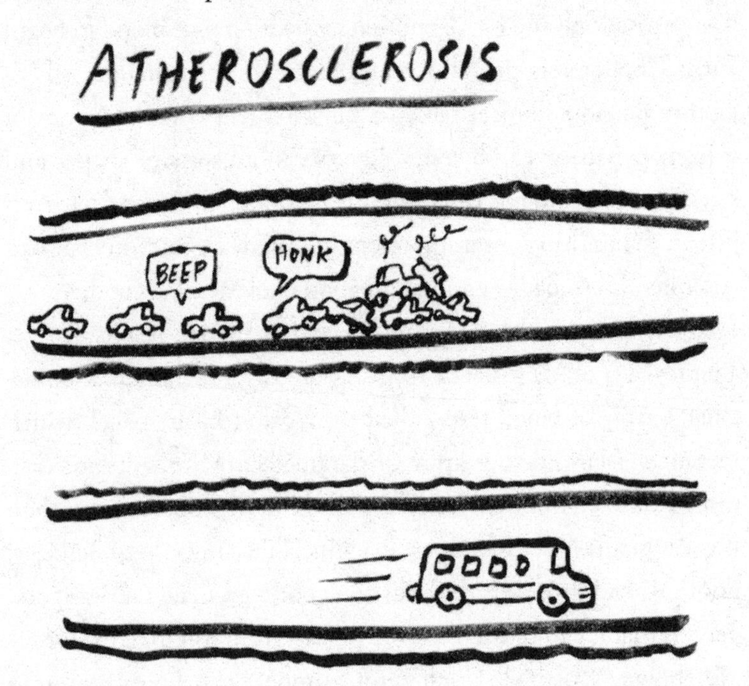

Each carrier molecule is marked by a protein called apolipoprotein B, or ApoB, which shows up on tests. Major studies going back nearly fifteen years have shown that ApoB is a much better predictor of risk than plain old LDL, and some researchers have suggested that doctors should focus more on ApoB instead. Such testing is now routine in Europe, but unfortunately, the latest U.S. cholesterol guidelines do not even mention it.

How important is particle size? One function of the CETP

"longevity gene" identified by Nir Barzilai in his centenarians is to increase their LDL particle size. Centenarian Irving Kahn's LDL cholesterol probably gets shipped around in particles the size of jumbo jets. Unfortunately, I don't have the right CETP gene, so my ApoB score was 101, a good bit above the risk threshold of 90. So, apparently, my LDL rolls around in rusty pickup trucks flying the Confederate flag.

Luckily my HDL cholesterol number was also high, which is good because one function of HDL is to "sweep" cholesterol out of the arteries and back to the liver. Better yet, most of my HDL particles seemed to be large, a type called A-1, which are the more effective sweepers. (Best of all: red wine is known to boost HDL.) Lebowitz scribbled on my sheet, "Great!"

Still, it wasn't *that* great: my HDL and LDL numbers would technically make me a candidate for "cholesterol lowering," he said, which usually means a statin drug. It's almost a rite of middle-aged passage, that moment when your doctor writes your first Lipitor prescription. But I was not eager to take a statin, especially with Nir Barzilai's words ringing in my ear. Mainstream cardiologists almost uniformly believe that their advantages greatly outweigh any downsides, but my latent Christian Science genes nevertheless objected to a long-term medication—not yet. Luckily, a little more digging proved that I didn't need one.

There are actually two kinds of cholesterol in your body, the stuff you produce and what you absorb from your diet. The latter actually makes up a relatively small portion of the total, yet our bodies are remarkably good at preserving and recycling it. "Remember, our genes still think it's three thousand years ago and we're starving, so these mechanisms for preserving cholesterol are still going strong," Lebowitz said.

The thing is, a statin drug would only reduce the cholesterol I produced, but the fancy test showed that my particular type of cholesterol was mostly absorbed from my diet (apparently, they can tell the difference). So a statin wouldn't do much. This was good news: I could use something less potent, a drug like Welchol or Zetia, which simply helps transport cholesterol out of the intestines. (Oat bran fiber does much the same thing, absorbing cholesterol in the gut and escorting it out of the body.) But I still demurred, so he began quizzing me about my eating habits.

"Do you eat a Mediterranean diet?" he asked.

"Sort of," I hemmed.

"Do you like French fries?" he prodded.

"Once a week," I lied.

"Noooooo—that's not acceptable."

"Steak once a week?" I ventured.

He kind of grimaced at that news, too.

Red meat has long been known to be a risk factor for heart disease, originally because of its fat content. More recent research has more or less exonerated fat, and suggesting that in fact the French fries and their carbohydrates may be far more dangerous to one's arterial health—contrary to three decades of received wisdom and government-approved dietary advice. In other words, eat fat, not bread. Recent research has suggested that dietary cholesterol may actually have a negligible impact on heart-disease risk.

So there must be other reasons why red meat does appear to increase cardiac risk, and some studies have identified a possible culprit in carnitine. Carnitine is metabolized to a chemical called TMAO that causes atherosclerosis (think: Tenderloining My Ass Off). In a recent small study, researchers pinpointed particular gut microbes in most people that are responsible for producing

TMAO—but longtime vegetarians, who don't have those particular microbes, also don't make TMAO when they eat red meat.

Translation: According to this study, it is apparently only safe to eat meat if you're a vegetarian.

Which I wasn't. Nor was I about to become one. But I wasn't up for taking medication, either, so I instead vowed to eliminate burgers and French fries from my diet. Mostly. I could have gone further, and cut out all processed meats—which have been shown, in at least one large recent study, to increase heart failure risk more than unprocessed red meats. But I like prosciutto too much. That was enough to convince Lebowitz to let me escape from his office (I'd been there a full hour), although I'm sure he knew that, like most cardiac patients with middling numbers, I wasn't quite ready to do anything to change—and possibly even save—my life.

Then I did still more research, and that's when I really got freaked out.

Heart disease is typically thought to be a purely modern ailment, a function of our newly rich diets; back in the good old days, before we ate French fries and trans fats and tenderloined our asses off all the time, it supposedly didn't exist. But that actually isn't quite true.

In 1909, a French British scientist in Cairo named Marc Ruffer dissected and analyzed a group of ancient Egyptian mummies to which he had finagled access. He styled himself a "paleopathologist," one who studied ancient diseases, and in these dried-up old corpses, he found several. Some of the mummies carried eggs of the waterborne parasite that causes bilharzia, while at least one appeared to have pustules consistent with smallpox. Most interesting of all, he also found that several of the mummies appeared to suffer from what was thought to be a brand-new condition: atherosclerosis.

At the time, heart disease was just emerging as a leading cause of death in the industrial world—which it has remained, for more than a century. For years, the presence of arterial disease in these ancient mummies was thought to be a function of their decadent royal lifestyle, since only kings and queens and high-status individuals got to be mummified. But a little over a century after Ruffer, a multinational team shattered that view by reporting in the *Lancet* that they had found similar degrees of arterial hardening not only in dozens more Egyptian mummies—even some who were only in their late twenties at death—but also in numerous other mummified corpses from Peru, the American Southwest, and the Aleutian Islands in Alaska. (Atherosclerosis also afflicted Otzi the "Iceman," who was entombed in a glacier 5,000 years ago.)

Unlike the Egyptian dead, these individuals were far from royalty; in Peru, especially, mummification of the dead was a common practice. "In the same way I have my portrait of my Newport banker grandfather in my house, they would keep the mummy around," says Caleb "Tuck" Finch, a USC gerontologist who was part of the *Lancet* team. "Some of them were two thousand years old." The hope was that someday, perhaps, mummified Great-Great-Great-Great-Grandma would magically come back to life, like Ted Williams's cryonically frozen head.

The crucial difference was that the Peruvian mummies, and also the North American ones, were largely hunter-gatherers, eating a "healthy" and authentically Paleo diet—yet they, too, suffered from the beginnings of cardiovascular disease. Atherosclerosis affected women and men, at all age brackets, and all levels of ancient societies, and so the *Lancet* study raised a troubling question: Are humans somehow hardwired for heart disease?

Short answer: Yes, we probably are. That explains why those poor dead young American soldiers had arteriosclerosis before they turned twenty-five. In a sense, we are almost programmed to develop a certain degree of hardening of the arteries—if we live long enough. And that's not all: Our hearts themselves stiffen and atrophy with age. That was what had happened to my dog Theo, his poor heart prematurely aged by the strain of feeding his blood-hungry tumor. But the thing is, it happens to all of us eventually.

The most fascinating moment of The Blast, and also the most unnerving, came on my second day, when I lay down on a gurney in a darkened office, while a technician lathered my chest with a greenish gel. Then he pressed the cold metal end of an ultrasound wand into the puddle of goop and turned his computer screen toward me. Up popped a ghostly green image of my beating heart, something I had never seen. It looked bizarre, the valves flopping and waving around in a complex, fluid dance, like some kind of deep-sea animal.

It was doubly weird because the ultrasound revealed only a slice of an image, as though he had shone a flashlight into a dark, watery cave where my heart happened to be beating. We sat there and watched my heart work through every beat, the valves fluttering and wobbling before they finally snapped into place and allowed the chamber to fill. It was beyond strange to watch the single muscle, heretofore invisible, that had sustained my entire life.

We were watching elasticity in action: The heart muscle must stay strong enough to swoosh blood throughout the 60,000 miles (!) of blood vessels in a typical human body, contracting and expanding up to 180 times per minute at peak exercise intensity. Yet the arteries themselves must also remain flexible enough to handle all that flow—like a water balloon, they can only take so many fillings before the rubber starts to get sketchy. Over time, we lose that flex. And not just because of atherosclerosis, but because our hearts—like the rest of us—are subject to damage from *intrinsic* aging, the same way the fuel pump in your old car will eventually wear out and need to be replaced. It just gets old. It was never meant to last that long.

Like that old pump, your heart performs less and less well with age. Although your resting heart rate doesn't change much, except in response to endurance training, maximum heart rate declines pretty much in a straight line with age. So does VO_2 max, as I learned in The Blast. The heart muscle weakens and can't throw out as much blood per stroke. It still lasts much longer than any part of any automobile ever manufactured—it's a miraculous little machine, in fact—but the truth is that the human heart only lasts so long. Which is why the biggest, most serious, and least modifiable "risk factor" for heart disease is age itself.

As he moved the chilly ultrasound wand around on my chest, the Blast technician would periodically stop and press a button on his computer, capturing images of certain parts of my heart. He was essentially taking its measurements, paying special attention to the left ventricle, one of the parts of the "pump" that first shows signs of age. Over time, thousands and millions of contractions, the muscle tends to become thickened and enlarged, which you might think was good news. Bigger, stronger heart = more blood pumped, right?

Wrong. Actually, a larger left ventricle is a classic symptom of cardiac disease and high blood pressure, as the pump has to work harder to keep things moving through your stiffening circulatory system. As it pumps harder, it makes more muscle, and thus it gets bigger—and a lot less efficient. That was Theo's problem. He was only able to seem healthy, despite his huge tumor, because he had been so athletic all his doggy life. Aging is hiding in our bodies.

Long-term aerobic exercise and/or strength training can actually reduce one's risk of cardiac hypertrophy—although exercise started later in life is much less effective. If you let it go, an enlarged heart can segue into heart failure, which is one of the main reasons why older people run out of gas, as Luigi Ferrucci noted. Says Richard Lee, a Harvard cardiologist and stem-cell researcher, "Nature probably cares more about hair loss in the forties than it does about heart failure in the eighties—but now our hospitals are filled with people who are in heart failure in their eighties."

What can be done? Controlling blood pressure in general seems to help, either with medication—or with meditation, which has actually been shown to work in studies, believe it or not. Data from the BLSA has also pointed to abdominal fat—my kind—as another possible cause or contributor to left ventricle problems. So getting rid of that would be a good next step (one we'll explore further in a few chapters). But those only take you so far. The fact remains that like the components of your car, your heart simply was not designed to last forever.

Neither were our arteries. Next, the technician applied the wand to the inside of my neck, and waggled it around a bit. Now he was measuring something called "intimal medial thickness," or the thickness of the wall of my carotid artery. Over time, our artery walls tend to grow thicker—and less flexible. Hence we get high blood pressure, and cardiac hypertrophy, in a vicious feedback loop. Dr. Lebowitz had done the same measurements and while he found no actual plaques, which was good, he nonetheless concluded that my "arterial age" was fifty-nine, which shocked me.

Over time, data from The Blast has shown that by far the biggest contributor to heart disease and heart failure is our own aging biology—the way our heart muscle and arterial walls themselves change over time, making us more susceptible to athero- and arterio-sclerosis. Processes inside our own cells create various kinds of junk, including so-called cross-linked proteins and excess deposits of things like calcium and collagen.

We actually *want* calcium and collagen in our bones and our skin—but in our hearts, they cause hardening and stiffening and other bad things. And there seems to be no way to avoid them: The longer we live, the more of this junk gets deposited in our most important muscle. Worse, that muscle does not really regenerate, because heart-muscle cells do not divide. (Nor do neurons, the other most important kind of cell in our bodies.) And as more of us live beyond the traditional heart-attack risk range of our fifties and sixties, more people are living with aged hearts.

"You're just not gonna get a Pinto to do four hundred thousand miles," says Lee of Harvard. "You may get a Volvo or a Subaru, but you're just not gonna get a Pinto to do that. And some parts of the heart are probably like that."

So the trick becomes, can you change a Pinto into a Volvo?

Chapter 7

BALDNESS AS METAPHOR

A moist eye, a dry hand, a yellow cheek, a white beard, a decreasing leg, an increasing belly…your voice broken, your wind short, your chin double, your wit single, and every part about you blasted with antiquity.

—Shakespeare

Somehow, I managed to escape Baltimore without having to fill the *Downton Abbey* hat. My microbiome will go unanalyzed, at least for the next three years. But the good people at The Blast also ignored one other biomarker of aging that, for me at least, seems all too obvious: my hair. More specifically, the lack thereof. It's the moment I most dread, when my hipster barber hands me the mirror to "take a look at the back" and the creeping bald spot that an LA standup comic once called a "flesh-colored yarmulke."

I'd been obsessing about my flesh-colored yarmulke since earlier in the summer, when I drove up to my college reunion (let's just call it a "significant" reunion). I was there partly out of curiosity about what kinds of adults my pimply classmates had turned into, but also for the usual, time-honored reason: to see how old everyone else looks. Over a June weekend, a few hundred of us congregated on Ye Olde College Greene to get reacquainted. We bunked in the old dorms, traipsed around to the old landmarks, and remarked on how our favorite professors now seemed so aged which is precisely what the students were thinking about us. In the evening we sipped rather than chugged our beers under white tents and tried to sustain conversations with people we hadn't seen in decades. One major question was always left unspoken: *What in the hell happened to your hair?*

Before I left, I had dug out the old "Freshman Facebook." I am so old that back then it was an actual, physical book, with a photo and basic information on every single member of the class of [REDACTED]. But what jumped off the page—almost literally—was the hair. We had so much of it, incredible, thick, dense masses of hair sprouting so far down our foreheads that it almost merged with our eyebrows. In many photos, it seemed as if some kind of furry animal had nested on the person's head.

Granted, those photos were taken in the 1980s, sort of a lost golden age for hair. But still: There was a lot of hair. And now, for most of us, it was at best a vestige of its former self. Many of the guys had surrendered to the inevitable and just buzzed their skulls; others, including yours truly, were still in denial. Not just the men, either, but many of the women had also lost the sheen and luster of their youthful tresses. But not all of us were so afflicted, at least not yet: A select few classmates still sprouted—nay, flaunted—their luxuriant manes, perhaps with a few gray flecks, but otherwise little changed from their freshman shots, except that the rest of us now hated them.

So what had happened? Why was our hair going bad, just two decades after its glorious, shaggy peak? And was this some sort of marker, a sign that the rest of our bodies were also dying, withering, falling apart?

Seeking answers, I went to visit Dr. George Cotsarelis, a professor at the University of Pennsylvania and a leader in the science of hair regeneration. It's not that our hair *disappears,* he explained, in a soothing voice practiced on thousands of panicked patients, who may or may not include Donald Trump. A grip-and-grin shot of Cotsarelis and Trump adorned the office bookshelf, right next to one with President Obama, but he would neither confirm nor deny that he was the genius behind The Donald's famous mane. I didn't press the issue. I had deeper concerns.

Hair loss is not really loss, Cotsarelis explained. It's more an issue of shrinkage. The good news is that our hair follicles have not disappeared, even in my flesh-colored-yarmulke zone. The bad news is that those follicles have "miniaturized" to the point where the hairs they produce are microscopic. In other words, bald guys aren't really bald; their hair is merely invisible. But beyond that, he didn't really have many answers. "We don't really know much about why it happens," he admitted. "We really don't."

You're not helping, I thought.

The medical term for male pattern baldness is *androgenetic alopecia*, and indeed it seems to be hardwired into the genes of many unfortunate men (actually, most men). But in fact, most of Cotsarelis's patients are women who suffer from basically the same syndrome: Their hair volume shrinks, and the strands get thinner, even if they don't eventually fall out or "miniaturize" like their husbands'. "It's interesting," he observed, "the ones who are most affected by this are the ones who had the most beautiful, dense hair when they were younger. They were in the ninety-ninth percentile."

Surprise: Hair loss, or hair weirdness, affects women, too. According to one study, 6 percent of women under age fifty show some degree of actual hair loss—a figure that rises to 38 percent by the time they reach their seventies. For men, of course, it's much worse: Four out of five will lose major hair by age seventy. But even among those who do keep their hair, nearly everyone will go gray eventually. Why is *that*? I asked.

Gray hair is not caused by individual hairs "turning" gray, Cotsarelis explained patiently. Rather, gray hairs are simply hair strands that lack pigment, another symptom of aging follicles. Over time, the pigmented hairs fall out, leaving the gray ones behind. If you're lucky. Only 5 percent of men or women keep their youthful-looking locks into their seventies, he estimates.

For the rest of us, Cotsarelis feels our pain: "Evolutionarily, I think the state of your hair was a very important indicator of health," he said. "If you see somebody with luxurious, thick hair, you know they are not nutritionally deficient, they have to have had enough calories—and they're probably fertile. But if they're sick, if they have mange, their hair's looking terrible, and they're not attractive. That's all evolutionarily built in."

For most of the Class of 198-, apart from a lucky and annoying few, our barren scalps were telegraphing to the world that we were tired, used-up, and perhaps close to the end of the line—bad bets for swapping genes. Which is why ladies often hesitate to click on bald dudes on Match.com, and why the hair-products industry takes in billions of dollars from both sexes. It isn't just vanity—it's *evolution*. "It's just an enormous part of our identity and sense of well-being," Cotsarelis said.

What had brought me here was the fact that Cotsarelis has done groundbreaking research into how hair follicles can be induced to regrow. In 2012, his team identified a key culprit in hair loss, a molecule called prostaglandin D2 that is often found around the scene of the crime: areas of the scalp that have no hair. The evidence is more than circumstantial: Prostaglandin D2 is known to inhibit hair follicle growth, and is also related to inflammation, which tends to increase in our bodies with age. Merck was testing a drug that inhibited PGD2—inhibiting the inhibitor, if you will—but withdrew it in 2013. Cotsarelis has founded his own start-up company called Follica that is working on its own prostaglandin inhibitor, among other treatments.

But the real action is with hair stem cells, which Cotsarelis had "basically become obsessed with" during graduate school. Back then, it was thought that we were born with a certain number of hair follicles, and that they died out over time; in fact, Cotsarelis says, follicular stem cells remain intact throughout our lives, but as we get older, they simply go dormant—something that's true of other kinds of stem cells, too, it turns out. So the question becomes, how do you reawaken the stem cells?

In 2007, Cotsarelis published a novel paper in *Nature* that described how he had administered a bunch of tiny jabs to the skin of mice, and waited to see what happened to the skin cells during healing. To his surprise, he found that the wounds triggered a cascade of growth factors that basically returned the skin cells to an embryonic-like state—turning them to stem cells, in effect—and in turn, induced them to generate brand-new hair follicles. Perhaps human scalps could do the same thing, if given the right combination of drugs and jabs. He's working on that, too, with his start-up.

So is there hope for my ever-expanding flesh-colored yarmulke? My frontal hairline, in retreat like the Confederate army at Gettysburg? He thinks so, even if he wasn't terribly forthcoming with the details just yet.

"But even if we figure out how to treat it, that doesn't mean we'll 'cure' it," he cautioned as I left. "I think what we're doing will end up with treatments, without necessarily understanding it."

Curing baldness is pretty far down the list of NIH grant-funding priorities; most scientists who study aging think hair loss is pretty much irrelevant to our actual, biological aging. Evolutionarily, though, the fact that so many of us lose our hair has lots to do with why all of us age. Cotsarelis was right about that.

Not long after Darwin proposed his theory of evolution, scientists began wondering how aging and death might fit into his framework. Why did we age? How had natural selection permitted aging to exist, since it is pretty much the opposite of "survival of the fittest"?

In 1891, the great German biologist August Weissmann took a stab at answering that question. He speculated that living things grow old and die in order to make room for the next generation, saving resources so that the young may survive. In his view, aging had been programmed into us for the good of the species, and that the old are meant to die and get out of the way. This theory has been enormously popular with students and everyone else under the age of twenty-five. But the notion that aging is somehow programmed has been hotly debated ever since, and most evolutionary biologists disagree with Weissmann.

For one thing, scientists have long believed evolution works on the level of the individual, not the group; genes are selected and passed on because they benefit the animal that carries them, enabling him or her to reproduce. The idea of group-based selection goes against that. For another, very few wild animals live long enough to die of old age; most perish much earlier from other causes, such as being eaten. Look at mice, for example. In the laboratory, with regular feeding and a nice cage full of cozy sawdust, a mouse will live about two years before it dies, generally of cancer. In the wild, though, mice rarely live longer than six months, and they generally perish in the mouth of a fox or, far more commonly, from cold.

So aging is not hugely relevant to the evolution of mice or any other animal, including humans. Our average hunter-gatherer ancestor lived to be somewhere around twenty-five years old, most likely dying from an infection or an accident, or an attack by a predator or fellow human. Only a select few lived into their sixties and seventies—and this, realized the insightful British geneticist and eventual Nobel Prize winner J. B. S. "Jack" Haldane, might actually help explain why we age the way we do.

Haldane had been studying a condition called Huntington's disease, which might possibly be the world's most horrible illness. Basically a very early-onset form of dementia, it starts with subtle changes in personality and balance, but then develops into a torturous, crippling affliction within a few years. Its sufferers almost seem to be dancing as they lose control of their bodies and begin writhing and jerking spasmodically. Perhaps the most famous Huntington's victim, folksinger Woody Guthrie, spent the last fifteen years of his life in mental institutions before he died at the tragic yet typical age of fifty-five.

What struck Haldane as odd was that Huntington's is actually an inherited condition, carried on a single gene. Not only that, but the Huntington's gene is dominant, meaning that even if only one parent has it, his or her children will have a fifty percent chance of developing the disease. According to the theory of natural selection, such a catastrophic genetic condition should have been bred out of existence long ago. But Huntington's has one other unique quality: Its symptoms do not appear until around age forty.

Haldane realized what this meant: Because Huntington's only manifests itself later in life, after its carriers would have had children, it had remained largely untouched by natural selection. By the time a person realized they had the gene, they would have already passed it on. Thus the Huntington's gene was able to survive because it resided in the "selection shadow," the post-reproductive period of life where the force of natural selection is drastically weakened.

One of Haldane's colleagues, the brilliant Peter Medawar—also a future Nobelist—saw the connection to aging. The selection shadow allows all sorts of harmful genes to flourish in later life, not just Huntington's but many other unpleasant traits that otherwise should have been erased by the purifying force of natural selection: the hardening of our arteries, the softening of our muscles, the wrinkling of our skin, the blossoming of our love handles, and the steady unraveling of our brains. Not to mention the dreaded flesh-colored yarmulke. Once we hit middle age, evolution pretty much takes its hands off the steering wheel and cracks open a beer.

So whatever purpose me and my college classmates' luxuriant, youthful manes might have served—plumage to attract mates, insulation for our brains, protection against injury or the sun—they became irrelevant after forty, because evolution no longer cared how good looking we were. Same for our vision, our knees, and our sexual hydraulics. As the geneticist Michael Rose put it, "The latter part of the life cycle [becomes] a genetic 'garbage can,'" which is also where our hair ends up.

The selection shadow explains a lot of other things, such as why women are much likelier to get breast cancer in their fifties than in their reproductively active twenties. Mothers who were genetically predisposed to early breast cancer would have had a harder time raising babies, and those babies would have been less likely to survive and pass on those early-breast-cancer genes. The bad news is that by the time we hit middle age, the genetic garbage can is overflowing and stinky. A decade's worth of research into yeast recently found that by eliminating some 238 genes, it was possible to increase the one-celled organisms' lifespan by as much as 60 percent. Or, to put it another way, yeast are born with 238 genes—at least—that are trying to kill them. So instead of longevity genes, our DNA may be packed with the very opposite.

Medawar saw aging as the accumulation of these harmful, unselected-for genes. But what if the very *same* genes that sculpted us into our magnificent twenty-year-old selves actually end up killing us in the long run? This was the insight of an American geneticist named George Williams, who speculated in a 1957 paper that certain genes that are helpful early in life could themselves have harmful or even dangerous effects later on. This phenomenon was later dubbed antagonistic pleiotropy (*pleiotropy* referring to a single gene with multiple functions). Moreover, Williams claimed that natural selection would actually *favor* such genes.

An interesting example of a pleiotropic gene turned up recently, and it has to do with the fact that white people get tan. Scientists at Oxford University found a gene variant in Caucasians that helps their pale skins resist damage from the sun's UV rays by temporarily darkening, but at the cost of increased the risk of testicular cancer. To pick another example, scientists puzzled over the high prevalence of the gene for hemochromatosis, a condition that causes dangerously high levels of toxic iron in the blood, leading to damage and disease in midlife. But then evidence emerged from the Middle Ages that men with that gene appeared to have greater resistance to bubonic plague. The hemochromatosis gene gave you better odds of surviving the plague, at the cost of poor health in middle age. To evolution, that's a no-brainer: The gene stays.

Aging is full of these kinds of trade-offs, between immediate survival and eventual longevity. The loser, usually, is longevity. In fact, evolution may actually have helped make our lifespans *shorter.*

In the early 1990s, a young researcher at UCSF named Cynthia Kenyon discovered a mutation that vastly increased the lifespan of our friend *C. elegans*, the millimeter-long worm that is so beloved by aging scientists. The mutant worms lacked a gene called daf-2 that governs metabolism—specifically, receptors for insulin-like growth factor, the worm equivalent of our IGF-1. Kenyon found that her daf-2 "knockout" worms lived twice as long as normal or "wild type" worms.

This was a stunning discovery, the first real evidence that aging could be slowed by deleting a single gene. Not only that, but her discovery showed that the insulin/IGF "pathway" played a central role in aging—and that, contrary to the belief of Dr. Life et al, the less of these growth factors you have, the better. Kenyon asserted that her worms had survived for the equivalent of 120 human years, a claim that made headlines and put her on the long list for the Nobel

Prize. No comparable "Grim Reaper gene," as Kenyon dubbed it, has yet been found in humans, but her discovery gave a tantalizing hint that tinkering with our own genes might extend lifespan.

Such gene therapy is still years away, and it remains doubtful that anyone could find a single gene with such a dramatic effect on human longevity. (For one thing, *C. elegans* has only 959 cells, while the human body has trillions.) But the discovery also shed light on the evolution of longevity itself. Common sense would dictate that the long-lived worms should have enjoyed some sort of evolutionarily advantage over their short-lived cousins. So why hadn't natural selection *already* eliminated the Grim Reaper gene?

A few years after Kenyon made her discovery, a Scottish-born scientist named Gordon Lithgow figured out why it had. He mixed normal worms and long-lived daf-2 knockouts in the same dish, to see what would happen. He was stunned by the results: Within just four generations, the long-lived worms were all but extinct. The reason, it turns out, was that the knockouts reproduced ever-so-slightly later than the normal wild-type worms. It didn't take long before the long-lived mutant worms had been outbred, outnumbered, and overwhelmed by their slutty wild cousins. Their extreme longevity proved to be no help at all.

Interestingly, this phenomenon has also been observed in long-lived humans; Nir Barzilai's Ashkenazi Jewish centenarians, for example, tended to have very few to no children, despite having married in the 1920s and 1930s, before the advent of birth control. Or if they did have kids, they did so later in life—another finding that has been associated with longer lifespan. (A side effect of this, of course, has been to make "longevity genes" even rarer.) So perhaps having kids really does shorten your lifespan.

Lithgow's worm showdown seemed to prove that natural selection pretty clearly favors fast breeders over the long-lived, for the same reason most towns have twenty McDonald's for every fancy French restaurant. Cheap and easy generally beats out long and slow. But it also raised other questions, such as: Why *do* some animals—and some people—live longer than others? Why do humans live eighty years, while mice live only two or three under the best conditions? Why has longevity evolved, at all?

As it turns out, there are actually two explanations: One has to do with sex, and the other with death.

Steven Austad hadn't thought much about aging until the day in 1982 when he met Possum #9. Before that, as his wife observed later, he had been concerned mostly with sex. Possum sex, but still.

He was camped out in the savanna of central Venezuela, helping a friend trap female possums for a study. Every month or so, they would trap all the local females, evaluate their health, and let them go. One day he recaptured Ms. #9, whom he had first trapped and banded just a few months earlier. Back then, she had been young and feisty, delivering a "solid bite," he later recalled. Now she was suffering from arthritis and nearly blind from cataracts. When he let her go, she wobbled off and bumped into a tree.

He saw this sort of thing happen over and over: Within the span of six months or so, healthy young animals "would just fall apart," he says. "They were aging incredibly quickly."

The story got even stranger a few years later, when Austad went to study a different bunch of possums who lived on a remote barrier island called Sapelo Island, off the coast of Georgia. The island possums were basically the same as their Venezuelan cousins, with one major difference: They lived much longer, as much as four years, versus just a year or two for the Venezuelans. The main reason had to do with predators: The jungle possums had lots, but the island possums had none, having been isolated from the mainland for more than five thousand years.

As a result, they were rather more relaxed. The first one Austad spotted was snoozing in the middle of the road. He ran up and grabbed it with his bare hands. He soon realized that he had arrived in a possum paradise. With no predators, the animals' lives were virtually stress-free, with nothing else to do but eat, sleep, and breed, which they did with gusto. Each Sapelo female produced two or three litters in her lifetime, versus just one per female in Venezuela. It was like a Club Med for possums.

Austad was not surprised that the island possums lived longer, but what struck him was that they also seemed to be aging more slowly. He examined the tendons in the animals' tails, and found that the Sapelo possums stayed limber for much longer than the mainland animals, a telltale marker of slower aging; their limbs and joints did not grow old and stiff nearly as quickly. Far from aging at some fixed rate, he realized, the Sapelo Island possums had somehow evolved to age more slowly than their jungle cousins. It was Nature's version of the class-reunion problem: Some animals get old in the blink of an eye, while others seem to stay youthful forever. But why?

Unlike most aging researchers, who stay largely confined to their labs, Austad is a bit of a wanderer. He spent the early part of his career as a field biologist in exotic places like Papua New Guinea, as well as Venezuela and the Georgia coast. He's spent a fair portion of his life sleeping in tents, but it was a lion named

Orville who helped instill his keen awareness of his position on the food chain.

One winter afternoon, Austad drove me to one of his favorite spots, the San Antonio Zoo, not far from downtown. As we strolled around the park, he shared stories and fun facts about monkeys, spiders, pythons, and tree kangaroos, an odd animal that he had studied in Papua New Guinea. Humans nowadays, he observed, age a lot like zoo animals—protected from predators and accidents, we live a lot longer than we ever did in the wild. Eventually we found ourselves outside the lion enclosure, where a young male started giving Austad a hard stare, as if Austad had been hitting on his girlfriend in a bar. Or perhaps he was just bored; it was a chilly day, and the zoo was nearly deserted. Whatever the reason, Austad seemed unnerved, and we moved on quickly.

"I don't like to make eye contact with them," he said. "Even on safari, when we drive through a group of lions, I'll sit in the middle of the vehicle, not looking out. People think I'm crazy."

Then he tells them about Orville. Long before he ever thought of attending graduate school and becoming a scientist, Austad had chanced into a job as an animal handler for Hollywood movies. His task was to make sure the lions acted their part properly—yawning on cue, for example—and more important, to keep them from biting the actors. You might think that such an occupation would require years of training and experience, but you would be wrong: At the time, his résumé included "semi-employed English major" and "New York City taxi driver."

One day, he was out walking Orville on the ranch outside Los Angeles where the animals lived, when into the lion's path strolled an unfortunate duck. The lion pounced on the duck, and Austad ordered him to drop it, backing up the command by whacking Orville with the metal chain that served as his leash. "One of the things about dealing with big animals like that is they have to do as you say *now*, the first time," he says. "It's not like with my dog, where I'll say 'Come! Come! Come?' "

Orville dropped the duck, as ordered, but then he went after Austad instead. In short order, Austad found himself on the ground, with his right leg in Orville's mouth. When he finally accepted the fact that he couldn't escape, he managed to calm himself down and assess the situation. The bad news was that he was in the process of being eaten by a lion. The good news, if you can call it that, was that Orville was eating slowly, almost thoughtfully. "The fact that he had my leg meant he wasn't eating anything else," he said. On the other hand, he figured Orville would get to the rest of him eventually. Luckily, though, the ranch bordered a road, and just then a carload of tourists happened to pull over to gawk at the animals. The tourists saw a lion on top of a man, and reported the situation to the ranch office.

Austad spent the next several weeks recovering in the local hospital, where he became a bit of a celebrity because the ranch owner's wife—actress Tippi Hedren, star of *The Birds*—would visit him there every day. Miraculously, he ended up keeping his leg, minus a little bit of his femur and a lot of his blood. He even went back to work, but it wasn't long before Orville charged him again, and it seemed like time to find a new line of work.

With the exception of Orville, Austad had always loved animals, ever since he was a boy growing up on an Indiana farm, so he headed back to school intending to study lions in the wild.

Instead, he got diverted to less glamorous and less toothy critters, like the possums. When he saw Possum #9 stumble off and bash into the tree, he realized that aging was the one big, unsolved problem that affected everything that had ever lived. And we understood almost nothing about it. "It's the big enchilada in biology," he says.

The tale of the two possums seemed to confirm, at least broadly, an intriguing new theory of aging. Dreamed up by a young British scientist named Thomas Kirkwood (who had studied mathematics, not biology), the "disposable soma" theory basically said that our bodies are mere vessels for our reproductive "germ" cells—our DNA—and thus only need to last long enough for us to reproduce. What happens afterward is of no concern. And because the germ cells take priority, our bodies only need to last as long as we're likely to survive in the wild. So it makes no sense for nature to build a possum that is capable of surviving ten or fifteen years, when it is only likely to live for two or three years.

"All our genomes need to do is invest enough in the body so that the body is in decent shape for as long as we're likely to live," Kirkwood told me. "But in nature it's a bad strategy to invest more than that, because what is the point? Nature is a dangerous place, so you don't need a body that can go on indefinitely in tip-top shape."

But for possums that are lucky enough to have been born on idyllic, predator-free Sapelo Island, it makes perfect sense to build a longer-lasting body. Over five thousand years, they evolved to live life on a more relaxed timetable, without so much pressure to hurry up and breed. The obvious question, to Austad, was how are they different?

Most other scientists focused on "model organisms" such as fruit flies, or nematode worms like *C. elegans*, or mice—"not just mice, but one *strain* of mouse," he says. These animals had one trait in common: They did not live very long, which was convenient for obtaining quick study results, but Austad doubted they could teach us much about human aging. "Rodents for the most part are pretty boring, from an aging perspective," Austad says. So he decided to look outside the usual lab animals, to try to identify other ways in which Nature herself had managed to defeat, delay, or reprogram the aging process in the wild. He started by trying to keep a colony of possums, but they proved difficult to raise under lab conditions. So he began looking for species that already lived longer than one would expect.

For most animals, lifespan correlates pretty well with size, with bigger animals tending to live longer, in general. (Chihuahuas, and other small but long-lived dogs, are an exception to this rule, thanks to centuries of manipulative breeding.) Using a measure called the longevity quotient, which compares actual with size-predicted longevity, Austad found that humans are actually pretty long-lived, relatively and absolutely: Very few other living things have been observed to live a hundred years or more. So extending human lifespan even further would be quite a difficult trick.

But a few creatures do manage to outlive us, starting with the famous Galapagos tortoises, some of which had survived the depredations of hungry nineteenth-century whaling crews and lived into the present day—more than 150 years. Lobsters were also well-known longevity champs, at fifty to one hundred years. That's youthful compared with certain clams that have been known to live hundreds of years. One particular Icelandic clam was recently found to be more than five hundred years old, meaning it had been around nearly since the time of Columbus. The clam, nicknamed Ming, was living happily in captivity until 2013, when researchers had the bright idea of trying to open her up. Lord knows why they did this—maybe to make chowder?—but it resulted in her untimely death at the estimated age of 507.

There were more surprising cases as well. Ancient handmade harpoon points were found in the carcass of a bowhead whale killed by Inupiat Eskimo hunters in Alaska in the 1990s. Previously, the whales had been thought to live "only" about 50 years, but this specimen turned out to be some 211 years old, based on analysis of changes in its eye lenses. Cold water may be good for longevity: Alaskan fishermen have also hauled in specimens of rockfish, relatives of red snapper and striped bass, in excess of a century old.

Austad lumps these long-lived species together into what he calls Methuselah's Zoo, creatures who display little or no evidence of aging—or as scientists say, negligible senescence. And while it's not easy to conduct aging studies on deep-ocean whales that live for two hundred years, one other very long-lived animal is both easily accessible and very numerous: bats. In the whole mammalian kingdom, Austad found, only nineteen species had a higher longevity quotient than humans, and eighteen of them were bats. In the wild, some bats have been known to live as long as forty-one years, giving them an LQ of 9.8, or double that of humans. (The nineteenth is something called a naked mole rat, a creature so bizarre that I don't really want to get into it yet.)

A few years ago, Austad and a colleague collected a bunch of bats from a colony that lived underneath a bridge in Texas, hoping to answer a simple question: How were they different from, say, mice? And how did this help them live longer?

The most obvious way the bats were different was that they were still alive at ages seven and beyond, when mice would long since have passed on. Another difference was that while mice crank out a litter of five to ten pups every thirty days, bats produce one offspring at a time, once per year. It makes sense: Tucked away in caves, with the ability to fly away from their few predators, bats have the luxury of reproducing slowly—much like the island possums, or humans.

But what is their "secret"? Is it their high-protein, low cholesterol, bug-based diet? All the exercise from flying around? All that sleep during the daytime? Probably not. Instead, Austad looked deeper, at where aging really resides: in the animals' cells. In a study, he and colleagues put a bunch of bat cells in a dish and sprinkled toxic chemicals on them, to gauge their powers of stress resistance. Then they did the same to mouse cells and human cells. The bat cells withstood stress much better than mouse or even human cells. Simply put, the longer-lived animals had hardier cells. So they lasted longer.

It all has to do with cellular maintenance, the internal housecleaning mechanisms in all our cells. In long-lived animals, Austad and others have found, these maintenance programs tend to be far better than in the cells of short-lived critters like mice. So their bodies are better looked-after, and thus last longer. It's as if you owned two cars, an expensive mint-condition Jaguar that you used on weekends and for special events, and a cheap, well-used Ford Focus for errands around town. You'd take the Jag to an experienced, specialized mechanic, hoping to get as many years as possible out of it, but you would probably take the Focus to Jiffy Lube, because it's cheap and easily replaceable. The same thing happens on the cellular level. The mice get Jiffy Lube maintenance, while the bats get the Jaguar treatment.

So now the question becomes: Is it possible, somehow, to make our own cells more like bat cells, and less like mouse cells? More bowhead whale than brook trout? More Jaguar than Ford Focus?

To answer that, we'll first have to understand how and why our cells themselves grow old—which we know they do thanks to a scrappy blue-collar kid from row-house Philadelphia, who took on and obliterated a fifty-year-old scientific myth that had been invented and propagated by an old French Nazi.

Chapter 8

THE LIVES OF OUR CELLS

You don't really believe all this bunkum, do you? This, this holier-than-thou dietary crap, the enema treatments, the mud packs, the sensory deprivation? What's it going to get us—another six months of eating grape mulch and psyllium seeds? Another year? We die anyway, all of us, even the exalted Dr. Kellogg—isn't that the truth?

—T. C. Boyle, *The Road to Wellville*

According to Google Maps, it is a mere 104 miles from the Golden Gate Bridge to Leonard Hayflick's home on the Sonoma coast. So you'd think it would take about two and a half hours to drive there, max. But in the real world, Hayflick himself insisted, the drive takes closer to four hours. "Google Maps will send you the *longest* way," he growled in an email, after he agreed to meet with me in March 2013.

Instead, he snail-mailed me a Xeroxed, hand-drawn, extremely detailed map showing the correct route to his house, covered with scribbled exhortations ("OBEY ALL SPEED LIMITS!!!"). Just the last twenty-seven miles would take a full hour, he said.

Yet still I doubted him, one of the most important scientists of the twentieth century. So have plenty of other people, apparently;

hence the map, which I soon discovered was accurate in all of its particulars. There really were "*speed traps everywhere!*" and it really did take an hour to drive the last leg, along winding coastal Highway 1. By the time I arrived at Hayflick's door, exactly four hours after crossing the Golden Gate, I was slightly dizzy from carsickness. Not to mention late.

"You were right about the drive!" I blurted. He replied with a grunt. By now, he's used to people not believing him, even when he *knows* he is right.

Nearly sixty years ago, young Len Hayflick was working in the lab of Philadelphia's Wistar Institute, putting his newly minted PhD to use in the trenches of cancer research. His important but unglamorous job was to produce and maintain clumps of living human cells, called cell cultures, for Wistar scientists to use in experiments. This sounds simple enough, but Hayflick kept running into a problem: Every so often, his cell colonies would die out. Either he wasn't feeding them properly, or the cells were becoming contaminated, or something else was happening that he couldn't yet diagnose. Whatever was going wrong, it was clearly his fault.

This he knew thanks to the work of Alexis Carrel, a celebrated French scientist who had essentially invented the discipline of cell culture. In his lab at Rockefeller University in New York City, Carrel had kept a strain of chicken-heart cells alive for decades, beginning in 1912. These were the most celebrated cells in the world; every year, the New York tabloids would celebrate their "birthday," with reporters and photographers paying visits to them in a dramatic glass-walled amphitheater that Carrel had specifically designed to accommodate the media.

No one dared question his work; after all, Carrel had won the Nobel Prize in 1912, for developing novel techniques for suturing blood vessels. He was the leading light of Rockefeller University, which was dripping with Standard Oil money (even today, his portrait still hangs in the foyer). In the 1930s, he upped the publicity quotient even farther by working with Charles Lindbergh to design a special pump to help with organ transplants, a stunt that got them both on the cover of *Time*. The two men also shared a love of eugenics, which Carrel advocated in a notorious 1935 book called *Man, The Unknown*. Meanwhile, the chicken cells were still living in 1943, when Carrel, a likely Nazi sympathizer, finally left them and returned to collaborationist Vichy France.

He died the following year, but his dogma lived on: Thanks to Carrel, everyone in the scientific world "knew" that living cells were essentially immortal—that is, they could divide forever. In his lab at Wistar, though, Hayflick began noticing an interesting phenomenon. At the time, he was using cells taken from human embryos, because unlike adult cells, fetal cells had not yet been exposed to contaminating viruses. But since abortion was not legal or common in the United States in the 1950s, fetal cells were difficult to come by. He had to tend them with special care. After a few months, though, they invariably died out. A check of his books showed that the cultures that failed were always the oldest ones.

He decided to forget about cancer and try to figure out why he couldn't keep his cells alive. Eventually he came up with what he calls "the dirty old man experiment." In one dish, he combined a batch of "young" female cells, which had only divided ten times, with an equal number of male cells that had doubled forty times—the dirty old men. A few weeks later, he checked the dishes and found that only female cells remained. The male cells were gone. So either something had killed only the male cells, or there was some other explanation. Like that the older cells were simply dying.

He knew that his results would upset one of the major apple carts of modern biology, so before he published, he knew he had to try to get buy-in from the established experts in the field—men like George Gey of Johns Hopkins, who a decade earlier had isolated a culture of cells from a young woman who had died from an aggressive form of cancer. Those cells, now known as HeLa from the name of the donor, Henrietta Lacks, had proved incredibly useful in cancer research (and are the subject of Rebecca Skloot's amazing book, *The Immortal Life of Henrietta Lacks*).

Hayflick sent samples of his fetal cells to George Gey and half a dozen other cell-culture poobahs, with instructions to call if and when they stopped dividing. "These guys are the major personalities in the field, using their techniques, which they think are the greatest," Hayflick recalls. "So when the phone started ringing, telling me that their cultures were gone, I'm thinking, if I'm gonna go down in flames, then I'm gonna go down in flames with some pretty good company."

Long story short, he had shown that Carrel had been completely wrong. Cells seemed to have a limited lifespan, after all. But his paper was rejected out of hand. The immortality of cultured cells, insisted one journal editor (employed at Rockefeller U, coincidentally), is "the largest fact to have come from tissue culture in the last fifty years."

Hayflick's paper eventually appeared in a small journal called *Experimental Cell Research*, in 1965. In it, he showed in painstaking detail how twenty-five different kinds of fetal cells *all* seemed to crap out at about their fiftieth round of cell division. Far from being immortal, he wrote, normal cells have a finite lifespan. Furthermore, he noted, cells taken from older donors had fewer doublings left in them. Their cells, like their selves, were aged. And the dogma was dead. "There is serious doubt," he wrote, "that the common interpretation of Carrel's experiment is valid."

Next came what Hayflick calls the "three phases of a new idea": "First you're an idiot; second, it's meaningless; third, it was obvious all along—and nobody gives you credit for it."

But Hayflick is nothing if not stubborn. His determination had already taken him from working-class southwest Philadelphia where he had grown up, and had discovered his love of science via the time-honored route of blowing things up in the basement with a Christmas chemistry set, to the nearby University of Pennsylvania, where he earned a BA and a PhD in molecular biology. And it gave him the courage to go into the study of aging, which was considered a scientific "dumping ground," as he puts it. "To have admitted you were working in the field of aging, in the '60s, was a recipe for professional suicide," he says.

Where others saw suicide, he saw opportunity. By 1975, he was in line to become the first director of the new National Institute on Aging. But then he found himself in the middle of a bizarre scandal, in which another arm of the NIH accused him of essentially stealing one of the cell cultures he had used in the "dirty old man" experiments, a line of cells called WI-38, which he and a colleague had created from lung tissue of a fetus that had been aborted in Sweden in 1962. WI-38 proved to be the most durable and useful cell line ever created: It was versatile, easy to grow, and "clean," free of viruses and other contaminants. It proved an ideal vehicle for manufacturing vaccines against all manner of diseases, from rabies to polio to hepatitis B. Merck and other major pharmaceutical firms used it to produce vaccines against measles, polio, smallpox, and rabies, among others. Hayflick provided samples of WI-38 to them and anyone else who requested it, in exchange for a small shipping and handling charge.

While the vaccines produced with WI-38 have saved countless lives, they also propelled Hayflick to the center of the abortion controversy, with religious hardliners (including the Vatican) objecting to the fact that the cell line had originated in the tissue of an aborted fetus. But their complaints were nothing compared to the wrath of a far larger, more powerful foe: the federal government, which accused him of basically absconding with federal property to create WI-38, and then using it for personal gain. Hayflick says he used about $100 worth of grant-funded supplies to help start his cell line, but insists that he never profited personally from WI-38, even as pharmaceutical companies made billions from the vaccines they manufactured with it.

The controversy cost him the NIA job and his faculty post at Stanford, which fired him unceremoniously. He "absconded" (his word) with a liquid-nitrogen tank containing his precious WI-38 cells, strapped into the backseat of the family station wagon, beside his kids, and drove across the Bay to Oakland, where for a while he supported his wife and five children on $104 a week in unemployment benefits. He ultimately ended up taking a far less prestigious faculty post in Florida. Hayflick battled the government for years before the case was eventually settled in 1982—shortly after Congress passed a law that permitted researchers and institutions to patent and profit from inventions created with government funding, as a result of which we now have what is known as the biotechnology industry.

Today, at eighty-five, Hayflick sits in his living room overlooking the Pacific, still as healthy and pugnacious as a prizefighter in his prime. The infamous nitrogen tank containing the original WI-38 cells resided in his garage until a few months ago, when he donated it for research. Those frozen cells are now more than fifty years old—even older than Carrel's fake chicken-heart cells were purported to be. Hayflick himself is also doing well for his age: He is sharp, lively, combative. "I have no pathologies," he says, a fact he attributes to genes from his mother, who passed away a few months earlier at the age of 106. Even in his eighties, he remains a battler, penning frequent letters to the editors of scientific journals, as well as longer opinion pieces attacking the antiaging industry and the research establishment. "I broke my ass for twenty years trying to get people to accept my ideas," he says. "It was not easy, I assure you."

His chance observation, that cells don't live forever, is now enshrined as the Hayflick limit, as universally accepted as Carrel's immortal-cell dogma was in its day. Hayflick's two papers, originally published in obscure journals, are now among the most-cited biology papers of the last fifty years. More important, though, was the implication of the Hayflick limit, which shaped the entire field of aging research.

He thinks Carrel had perpetrated something close to fraud with his "immortal" chicken cells for it later turned out that Carrel's assistants had been replenishing them, inadvertently or not. But Carrel's wrongheaded ideas also had influenced the study of aging. To be blunt, Carrel didn't believe that aging was real. Rather than an inescapable reality, he wrote in 1911, aging was a 'contingent phenomenon." Given the right conditions, he asserted, he could keep a human head alive forever, just as easily as he had kept the chicken-heart cells growing.

"Senility and death of tissues are not a necessary phenomenon," he wrote; aging results from accidents and causes outside the cell, he insisted. And for decades, many scientists believed this, even after Hayflick published his two papers. During the 1950s, it was thought that aging was caused primarily by radiation from the sun and from nuclear activity (this was the Cold War, after all).

Hayflick's work showed that the aging process itself had to originate somewhere *inside* the cell itself. The implications for aging biology were huge. Our cells themselves grow old, are mortal. "I think of Hayflick's work as a huge turning point, because it focused attention on the possibility of studying aging at the cellular level," says Steven Austad.

It was indeed a defining moment in human biology, but it was also the point at which Hayflick parted ways with many of his colleagues. To Hayflick, his limit was essentially proof that *nothing* could be done to slow or stop the process of aging—that it was a natural, inevitable consequence of the fact that our cells also aged and died. "Interfering with the aging process?" he scoffs, toward the end of my visit. "That's the worst thing you could do. Have you ever thought that through? How long do you want Hitler to live?"

Fortunately, not everyone saw things his way.

Hayflick's papers had left two important questions unanswered: *Why* is there a Hayflick limit? And what, exactly, is its relationship to aging?

He himself remained deeply puzzled by one odd observation: His cells seemed to know how old they were. If he froze a batch of

WI-38 cells at, say, their thirtieth division, and then unfroze them a few weeks or months or even years later, they would resume dividing—but only for another twenty times. "They remember," he told me, still sounding faintly amazed.

There had to be some sort of counting mechanism, he finally decided. And it was clearly independent of clock time, as his experiments with freezing and thawing the cells had shown. A cell's biological age, therefore, had almost nothing to do with its chronological age. The only thing that seemed to matter was how many times it had divided. He and his students would spend the next decade looking for this counter, which he called the "replicometer," with no luck. It took another quarter century for the answer to emerge, and it came from an unlikely source: pond scum.

In the late 1970s, a young Berkeley scientist named Elizabeth Blackburn was looking at a simple but unique protozoan called *Tetrahymena*, often found in stagnant water (which is why she likes to call it pond scum). Blackburn noticed that *Tetrahymena* had lots and lots of repeating DNA sequences on the ends of its chromosomes. The sequences first appeared to be "junk" DNA, without any function, just two thymines and four guanines—TTGGGG—repeated many times.

These telomeres, as they were called, cap the ends of chromosomes, protecting them in a way that's often compared to the plastic tips on the ends of shoelaces. Telomeres contain no meaningful genetic information, just a repeating series of amino acids (in humans, the sequence is TTAGGG, slightly different from pond scum telomeres). But they are far from useless: They serve as a sort of sacrificial barrier, protecting the more important, information-carrying DNA as it is copied. With each successive cell division, the telomere "caps" are chipped away slightly. When they are gone, the "laces" themselves—the important DNA—begin to fray, and when the damage gets bad enough, the cell may stop dividing.

But as usual in science, one discovery merely led to more questions. If our telomeres eroded like this, then why were we still here? Our cells had to have some way to repair their own telomeres and keep their DNA intact.

A decade later, still working on the pond scum, Blackburn and her graduate student Carol Greider discovered the answer: an enzyme called telomerase, whose job was basically to repair the ends of chromosomes, tacking on more TTAGGGs even as they were chewed up with each successive cell division. Telomerase helped maintain the "caps" of our DNA, keeping the shoelaces from coming unraveled.

It wasn't hard to find correlations between telomere length and health. One major study conducted over a period of seventeen years found a strong association between telomere length and overall mortality. Not to be alarmist, but the shorter your telomeres, the shorter your life, the study found. In another, more revealing study, a UCSF colleague of Blackburn's named Elissa Epel studied a group of mothers who had cared for a chronically ill child for a period of several years—in other words, as stressed a group of people as you're likely to find. She found that the longer a woman had been in the caregiver role, the shorter her telomeres tended to be—the equivalent of between nine and seventeen years of additional cellular aging. Caring for aged parents would also tend to have the same effect—proving yet again that aging fuels itself.

Other studies found links between shorter telomeres in white blood cells and many common diseases of aging, or risk factors for them, including vascular dementia, cardiovascular disease, cancer, arthritis, diabetes, insulin resistance, obesity, and on and on. Endurance athletes, on the other hand, seemed to have rather long telomeres, relative to the average person. And some long-lived seabirds have telomeres that actually grow longer with time.

So, pretty clearly, people with shortened telomeres are messed up. But the studies left a major question unanswered: Are short telomeres a *cause* of aging—or are they merely a symptom of some underlying biological stress, from a psychological situation or a chronic disease? More recently, another large study of more than 4,500 people found that, if you control for unhealthy behaviors like smoking and alcohol abuse, there is no link between shorter telomeres and mortality.

Blackburn, Greider, and another researcher named Jack Szostak would eventually share the 2009 Nobel Prize for their discovery of telomerase. But it remains far from clear whether telomerase is a magic bullet for aging. Some evidence hints that it might be: In a widely publicized study published in *Nature* in 2010, researcher Ronald DePinho took mice that had their telomerase gene knocked out, and were in horrible health as a result, and gave them a telomerase activator. Their health was magically restored—which was a big deal because DePinho, now head of the MD Anderson Cancer Center in Houston, had previously been known as a telomere skeptic. In humans, studies found that people with low levels of telomerase had higher levels of six major cardiovascular risk factors. But critics pointed out that all the study had really shown is that it's really bad to have *no* telomerase.

The notion that our cells have a built-in "clock" that can be reset, with a simple enzyme, is immensely appealing because it's so simple. Why not simply add (or turn on) telomerase, and keep those cells dividing? Anti-aging doctors like Jeffry Life offer telomere-length blood tests, costing from $200 to nearly $1,000, that purport to measure one's cellular age. Those same anti-aging doctors also sell a purported "telomerase activator" called TA-65—based on "Nobel Prize Technology," according to its marketing materials—so long as you're willing to pay $600 for a month's supply. Not an issue for the likes of Suzanne Somers, who takes it, but the rest of us need to know that its active ingredient is derived from the Chinese herb astragalus, which is available from the Vitamin Shoppe for about $15 a bottle.

There's another problem, too: Activating telomerase might cause cancer. One thing that nearly all cancer cells have in common is amped-up telomerase. To repeat: Telomerase is activated in roughly 90 percent of tumor cells. Cancer cells have long telomeres, too, obviously (which is why they keep dividing), and in fact one focus of recent cancer research has been to find ways to *inhibit* telomerase in cancer cells. A mouse study of TA-65, sponsored by the manufacturer itself, found that it not only failed to increase their lifespan, but the mice who were taking the stuff actually developed slightly *more* liver tumors than the control mice.

"[Telomerase] is the single most distinguishing characteristic between cancer cells and normal cells," Hayflick scoffs. "So that should be a red flag. Would *you* let yourself be inoculated with telomerase?"

Hmm, not if you put it that way. Although at the same time, *short* telomerase also lead to cancer. More questions about the whole telomere/telomerase theory of aging are raised by the fact that some animals with very long telomeres and lots of telomerase actually live a very *short* time—such as laboratory mice.

At best, then, the jury is out as to whether short telomeres are truly the cause of aging—or, rather, a symptom of age-related diseases. What may be more important, anyway, is the *fate* of our cells, and what happens when they stop dividing.

One of the most important tests I was subjected to in The Blast was a simple blood analysis that can predict, perhaps more than any other single marker, the state of a person's health. It's also a test that your doctor will probably never give you. Certainly, Blast staff never talked about it with me or shared my results; I didn't even know it existed until weeks later, when I talked with Luigi Ferrucci.

The test detects something called interleukin-6, or IL-6, which is a kind of "cytokine," a chemical messenger produced by our cells. Normally, IL-6 is supposed to help fight off infections and heal wounds, which it does as part of the body's inflammatory response. But in older people, IL-6 and other inflammatory cytokines seem to be hanging around all the time, in ever-higher levels, for no apparent reason. It's one of the biggest mysteries in aging: The older we get, the more inflammation we carry around in our bodies, and nobody quite knows why. Where does it come from?

IL-6 is like the Lance Armstrong of the inflammatory cytokines, the leader of a dirty bunch. It is responsible for most fevers (one of its functions is to raise body temperature), but it also appears to control the release of dozens of other inflammatory agents, the way Lance once led the Tour de France pack. Oh, and it's deadly—or at least, it correlates directly with mortality rates. According to the twenty-five-year-long Rancho Bernardo study of older Californians, the higher your levels of IL-6, the earlier your checkout time from Hotel Earth.

It's also one of the markers to which Luigi Ferrucci pays closest attention in The Blast. Subjects with higher levels of IL-6 are more likely to have more things wrong with them—multiple diseases of aging, or other risk factors for death. "While we can't say there is a causal mechanism, this is one of the strongest biomarkers that we have," he told me.

In particular, chronic inflammation seems to greatly raise the risk of death from cardiovascular disease, cancer, and liver disease. Which makes sense: Inflammation helps to form arterial plaques, and constant exposure to IL-6 makes cells more likely to turn cancerous. Inflammation has even been implicated as a contributor to depression. As we get older, it becomes so common that one of Ferrucci's Italian colleagues coined the term *inflammaging* to describe the conjuction of the two. But nobody could come up with a satisfactory explanation why so many older people seem to suffer from this kind of low-grade inflammation, until relatively recently. And one possible answer, it turned out, goes back to Hayflick and his limit.

Hayflick recognized two possible fates for our cells: Either they become cancer, that is to say immortal; or they enter a state he termed replicative senescence. But what did the senescent cells do?

In the late 1990s, a cancer researcher at the Lawrence Berkeley National Laboratory named Judith Campisi began to look at that question. It had been thought that senescent cells were basically benign, sitting there quietly like nice old retirees at the local McDonald's. Campisi wasn't so sure. She also wasn't convinced that the Hayflick limit really "caused" aging, in any meaningful way. "You can go into a ninety-year-old person, and take a biopsy, and you get a lot of cells that are still dividing," she says. "So the idea that you got old and died because your cells ran out of cell division just didn't cut it for me."

She began to take a closer look at the so-called senescent cells, and found that they were far from the benign cellular retirees Hayflick and everyone else had believed them to be. Rather than just sitting there harmlessly, she found, the senescent cells oozed a brew of inflammatory cytokines. "The big *aha* came when we realized that when a cell becomes senescent, it starts to secrete molecules that cause chronic inflammation," she says. "And inflammation causes or is a major contributor to virtually every major age-related disease that we know of."

Senescent cells make very bad neighbors, less like those nice, McLatte-sipping retired folk and more like a Clint Eastwood character gone bad, sitting on his porch with a Budweiser, a lit cigarette, and a shotgun. Their toxic secretions help poison the cells around them, in turn making them more likely to become diseased or cancerous—or to go senescent themselves; senescence seems to be contagious. Senescent cells may also be why aging lungs seem to get stiffer: the cells and their secretions are the enemies of elasticity. The good news is that in living tissues, senescent cells are not all that common—the highest percentage ever observed is 15 percent (in the skin of very old baboons). But like Neighbor Clint, it doesn't take many of them to make the neighborhood an unpleasant place.

Born in Queens and still very much a New Yorker, with her

penumbra of frizzy brown hair, Campisi seems a bit out of place in her own office, in the sleek, I. M. Pei–designed headquarters of the Buck Institute for Research on Aging, a stunning postmodern marble palace nestled against a Marin County hillside. In 2005, Campisi and her colleagues found that most types of senescent cells had a typical "signature" of cytokines that they secreted, with IL-6 usually at the front of the pack. She dubbed this the senescence-associated secretory phenotype, or SASP, which is scientist-speak for a polluted cellular environment. (Scientists love acronyms, as you may have noticed.) Curiously, though, it was the same bunch of cytokines that are responsible for the basic low-grade inflammation that afflicts older people—which made Campisi and others wonder if senescent cells and SASP were helping to promote the aging process itself.

"[Senescence] evolved to suppress cancer, but we think what it also does is drive these degenerative diseases later in life—and it even will drive, we think, secondary cancer, late-life cancers, the cancers you get after the age of fifty," she says. "It drives cancer, it drives neurodegeneration, it drives sarcopenia [loss of muscle]. That's what senescent cells do: They create this chronic inflammation."

It's the catch-22 of aging: Cells go senescent instead of turning cancerous, but senescent cells, in turn, create inflammation that helps cause *other* cells to become cancerous. Yet senescent cells do perform one very important function: They help with healing. If you jab a mouse with a scalpel (or yourself, for that matter), some cells around the wound will immediately go senescent and start SASPing all over the place. That in turn helps heal the injury and protect from infection. So in the short term, senescent cells are essential to keeping body and soul together; but in the long run, they might kill you.

Perhaps the strongest evidence that cellular senescence hastens the aging process came from cancer survivors, who had been subjected to powerful chemotherapy drugs that left them riddled with senescent cells, thanks to rampant DNA damage that basically stopped their cells from dividing. In follow-up studies, researchers began noticing that these long-term cancer patients were developing other age-related diseases much earlier than normal. "Twenty years down the line, they are showing up in the clinics with multiple age-related pathologies, including secondary cancers that are unrelated to their primary cancer," Campisi says.

A similar phenomenon was observed in patients who had been treated for HIV with powerful antiretroviral drugs—which also left their bodies, and primarily their immune systems, junked up with senescent cells. Many of those former HIV patients have also been found to suffer from conditions such as atherosclerosis, which is fostered by high levels of inflammation. Because of all their senescent cells, these former cancer and HIV patients are basically bathing in inflammation, which may cause them to age more quickly. So what happens if you make senescent cells go away?

At the Mayo Clinic in Rochester, Minnesota, two thousand miles away from the Buck, a team of researchers staged an elaborate experiment designed to see what would happen if we could somehow flush the senescent cells out of an animal. It was anything but simple: For starters, lead scientist Darren Baker and his team had to create an enormously complicated genetically engineered mouse. They began with a specially designed mouse that lacked the gene for a key protein, and thus aged prematurely due to a pileup of senescent and dysfunctional cells. They then crossed that mouse with another one that they had created, whose senescent cells could be cleared using a special senescent-cell-zapping drug. (Told you it was complicated.)

The new, hybrid mouse was not only one of the most exotic and expensive rodents ever to walk the earth, but also one of the most unhealthy. They aged extremely rapidly (because of all those senescent cells), forming cataracts at an early age, and losing muscle and fat tissue like very old people who are basically wasting away. In short, they suffered from something like frailty, and they were only in mouse middle age. Plus, they got all shriveled and wrinkly. But when Baker's team used another special drug to clear away their senescent cells, their condition improved drastically: They became much stronger, and lasted longer on treadmill tests, the cataracts cleared up, and even their wrinkles went away. They were rejuvenated, in short—all without taking any mouse growth-hormone shots.

"Someday, you might go in and get your senescent cells removed, like changing the oil in your car," study coauthor James Kirkland told me. First, though, we'll have to figure out how to identify them, and then remove them without damaging their neighbors—no easy task, since senescent cells make up only a small percentage of cells overall, and are scattered all over the place. "It's still science fiction," Kirkland warns.

The job gets easier, though, if you can figure out where senescent cells tend to hang out—remember, they are relatively scarce. Kirkland, like Campisi, has been studying cellular senescence for years—and he believes that these cells are driving much of what we recognize as aging. The big mystery—one of the mysteries—was where, exactly, they could be found, as they are extremely difficult to tag and locate (except in million-dollar exotic engineered mice, that is). Kirkland concluded that the most powerful and malign pockets of senescent cells are found in one particular kind of human tissue, one that most of us find a bit too abundant: fat.

Chapter 9

PHIL VS. FAT

There is no love sincerer than the love of food.

—George Bernard Shaw

Phil Bruno was SuperSizing again. It was just past 5:30 p.m. on a February evening in 2004, and he was driving home from work. A few miles from his house, he pulled into a White Castle, one of the many fast-food outlets lining the Manchester Road in the suburbs of St. Louis. Sure, he was only a quick drive from his own kitchen, where his wife, Susan, was cooking their usual big Italian dinner, but he was hungry *now*. He'd been doing it so long, it was almost automatic.

Ten minutes later, with a bag of hot burgers on the seat beside him, he pulled into *another* drive-through—this time McDonald's. There, he ordered another meal, a Double Quarter Pounder with Cheese combo, plus an apple pie for dessert. Oh, and a chocolate shake to wash it all down. "The reason I did this is because I would be embarrassed to order too much from one drive-through," he explained to me in a matter-of-fact email. "I didn't want the person at the window to look at me funny."

Because when a man who weighs nearly a quarter-ton pulls up

and orders multiple Value Meals, that's the kind of thing fast-food workers tend to talk about. Particularly if the person has been doing it every single day, as Phil Bruno was, at that point in his life. And it didn't end at the drive-through: Some nights he would fix himself a quick sandwich while waiting for dinner.

Phil had always loved food; it was part of the fabric of his tight-knit Sicilian family. Grandma and her lasagna were right down the street. In college he had played basketball, carrying 215 pounds on his big-boned, six-three frame. But then he'd gotten married, and had two children, then three. Suddenly, his weekends and evenings became all about the kids: homework, dinner, baseball and soccer practices. He stopped exercising. He didn't mind; everything was about family. But the weight piled on, pound by pound, year by year, without stopping. He didn't fight it. He even embraced it. Phil Bruno does everything with gusto, and what he did, at that point in his life, was eat. "I had to work *hard* to get this big," he says now.

He didn't have to work that hard. In part, Phil Bruno was a victim of his own middle-aged biology. As we grow older, and growth hormone and testosterone decline (along with other chemical changes), the calories we consume are far more likely to end up as fat. Beginning around age thirty-five, our total body fat percentage increases by as much as 1 point per year, even if our overall weight stays the same. More important, the distribution of that fat changes, from "subcutaneous" fat—the fat under the skin that makes young people look ripe and smooth—to abdominal or "visceral" fat, also known as a gut or a belly, or that stupid little pooch that you just can't get rid of. Waist circumference also expands, seemingly unstoppably, increasing about an inch and a half every nine years, according to one long-term study of middle-aged women. These changes are almost universal: If you compare an older person and a younger person of the same weight, says Luigi Ferrucci of the NIA, the older person will almost always have more visceral fat on them. We'll be able to survive longer when the Apocalypse comes, but we won't be particularly good looking. As an older but athletic friend of mine recently put it, "Sometime in your mid-fifties, everything just *changes shape*."

This I knew all too well, thanks to The Blast. A couple of months after my stay at Harbor Hospital, a manila envelope arrived in the mail containing my results. They were mostly unremarkable, except for one thing: my body-fat percentage was a lifetime-high 24.3 percent. My body was nearly one-quarter fat, which put me on the very high end of "normal"—and right up against the threshold for obesity, which is 25 percent body fat and above. For fit males, normal is more like 17 to 20 percent; for athletes, which is what I used to consider myself, it's lower, like from 13 percent down to 6 where things start to get body-builder freakish.

Women tend to run a few points higher than men, because their bodies are built to pack on fat for nurturing babies, so normal is more like 25 percent—and obesity starts at 32 percent body fat. Which means that some very obese women can be nearly one-half fat tissue. On the other hand, this means that women are more likely to survive a famine than men, which is probably the point.

Even though I was still within the realm of "average" (barely), I did not handle this news very well. I emailed the Blast staff straightaway, informing them that their machine was miscalibrated. Curtly, they informed me that it was not: It was a dual-energy X-ray absorptiometry, or DEXA, scan machine, far more accurate than the old-school method of measuring fat that involved a creepy guy with calipers and a clipboard. Unfortunately, their results also agreed with my $100 bathroom body-fat scale, which had recently shown a jump from a still-acceptable 18 percent to my current, porcine 24 percent. The nice lady informed me, helpfully, that according to the body scan, most of the fat was found in my midsection. "Welcome to middle age," she wrote. So that was that.

Phil was on another level altogether. My body-mass index, or BMI, was a shade over twenty-five, on the low end of the overweight range. Anything above thirty is considered obese. Phil's BMI was an off-the-charts forty-five, but more importantly, his waist circumference was nearly equal to his height; studies have shown that one's waist size should be less than half of one's height.

And it was more than just a number. All his eating had turned him into a "physical and emotional wreck," as he describes himself. His joints ached whenever he had to go up and down the stairs in their two-story home; his legs "felt like they were filled with sand." His heart hammered away inside his chest, all the time, and he was possessed by a strange, burning thirst that no amount of ice water could quench; all it did was send him staggering into the bathroom every half hour, all night long. "I was forty-seven years old," he says, "but I felt like I was eighty."

As it turned out, though, middle-aged biology was no match for Phil Bruno.

Prodded by a friend, he finally went to see his longtime family physician, whose name was Dr. Ron Livingston, on June 6, 2004. He'll always remember the date, because the results were so sobering. For starters, he couldn't even be weighed on the office scale, which maxed out at 350 pounds. He had to go to a nearby grocery store, and step on the scale used to weigh pallets of food off the delivery trucks. It said he weighed 475 pounds. Phil's blood pressure was at a fire-hose-like 250 over 160, putting huge strain on his arteries and his heart. His blood sugar was also off the charts at 600, or six times higher than what is considered normal; and his A1C, an important blood marker for diabetes, which should have been under 5.8, was a sky-high 16.

He had full-blown diabetes, obviously, but that was just one of his problems. He walked out of the doctor's office with prescriptions for no less than twelve different medications and supplements, from fish oil to blood-pressure medicine to Lipitor for his cholesterol to Glucophage for his diabetes. And he never forgot what Dr. Livingston had said, when he finished checking him out: "Bruno, I'm surprised you're even still alive. I'm expecting you to drop dead right here in my office."

Everyone "knows" that being fat is bad for you, but most people would be at a loss to explain exactly why. Some reasons, of course, are obvious: More weight means more stress on your joints and, more seriously, on your heart. And it tends to go hand in hand with diabetes, which is what Dr. Livingston had diagnosed Bruno with that day. Although not all heavy people are diabetic, most people with diabetes are overweight or obese.

Diabetes itself is now thought to speed up the aging process enormously—which is why diabetics generally give up about five to seven years of life expectancy. The body becomes unable to process the sugar that we eat, which ends up rollicking around in our bloodstream, inflicting massive amounts of cellular damage in every tissue that it touches. Excess blood sugar even makes you *look* older: One study showed that people with higher blood sugar actually did appear older than they were, perhaps because this damage is visible in their very skin. The older we get, the less efficiently we process sugar, and the more prone to diabetes we become. Centenarians, on the other hand, seem to be able to handle sugar with no problem—like my grandmother, who scarfed down a breakfast pastry every morning for her entire life with complete impunity. Glucose is her bitch.

But diabetes is only part of the reason why excess fat has also been associated with serious health problems including cancers of the kidneys, colon, and liver. A massive 2003 study published in the *New England Journal of Medicine* found that the high rate of obesity in the United States—one-third of the population—is responsible for 14 percent of cancer deaths in men, and 20 percent in women. New evidence points to the possibility that fat itself may be causing all these problems.

Until fairly recently, fat tissue was thought to be inert, mere energy storage for the body—as passive as a passbook savings account. You "deposit" calories by eating, and "withdraw" them by exercising. If you burn thirty-five hundred calories by jogging (for about five hours), then you will lose one pound, or at least not gain it. Otherwise, the fat is sitting there, not doing much of anything, or so it was thought.

In the 1990s, scientists began realizing that our blubber might do lots more than just jiggle. Over the last decade, they have come to recognize that fat is in fact a huge endocrine gland, and it wields powerful influence over the rest of the body. "For a typical North American, their fat tissue is their biggest organ," says James Kirkland, who helped pioneer the study of fat's endocrine effects. Kirkland believes that when it comes to aging, fat could also be the body's most important organ.

It's remarkably easy to gain weight—as both Phil and I learned—even huge amounts of weight, without really trying. Conditioned by millennia of feast-and-famine cycles, the human body has evolved into a remarkably efficient fat-storage machine, whose mission is to squirrel away any precious excess calories.

Evolution has not yet caught up to the fact that food is now relatively abundant and cheap for most of us. Our genes still think we are hunter-gatherers, and to them, fat equals survival. A massive study in the *Lancet* estimated that taking in just ten more calories than you burn every day can lead to a significant weight gain of twenty pounds over twenty years. Go from 10 extra calories to, say, 138—the 12-ounce can of Coke I used to drink every afternoon—or a 200-calorie candy bar, and those extra twenty pounds can pretty quickly turn into fifty or one hundred.

But not all fat is bad. Subcutaneous fat helps protect the body from injury, like padding, and it also secretes immune factors that help fight infection and heal wounds; that's where those wound-healing senescent cells come from. Fat in general is highly resistant to infection, and some scientists think it is crucial to the functioning of the immune system as a whole. It also keeps us warm in cold weather, keeps us afloat in water, and it can look nice, if properly placed. Without it, there would be no Kardashians.

"SubQ" fat also produces a hormone called adiponectin, which appears to help control metabolism and to protect against certain cancers, notably breast cancer—plus other good things that have not yet been identified. It is no accident that Nir Barzilai's Jewish centenarians tend to have higher-than-normal levels of adiponectin.

So much for the good news. The bad news is that, as we age, we gradually lose this good fat, which is one reason why our hands get more bony looking and "interesting." Instead, we pile plump, juicy fat onto our midsections, forcing us to buy ever-larger pants. This "visceral" fat is not the same stuff as the nice, wound-healing, adiponectin-secreting subcutaneous fat. For example, subQ fat also produces leptin, another important hormone that tells the brain, *Hello, you have plenty of stored energy, so you can stop eating now.*

Visceral fat produces very little leptin, so the satiety center of the brain never gets that message. This might be because visceral fat served a different evolutionary purpose, as short-term energy storage designed for quick access to produce short bursts of energy, like during an intense hunting session (which may be why men have more visceral fat, and women more of the "nurturing" subQ stuff). The stress hormone cortisol, activated during fight-or-flight situations, tells the body to sock away still more visceral fat—which, if your stress comes from a sedentary desk job, you never really get to burn off. Instead, the visceral fat just sits there, in between your liver and other vital organs, while you're stuck at your desk shopping online for new pants.

The pants aren't the problem. The fat is the problem. Over the last decade or so, and especially since Phil Bruno was diagnosed in 2004, Kirkland and other scientists have discovered that this abdominal or visceral fat infiltrates our vital organs, bathing them in a nasty chemical stew that wreaks havoc throughout the body. Visceral fat produces an array of inflammatory cytokines including not only IL-6, the king of chronic inflammation, but another one called TNF-alpha, for "tumor necrosis factor," which is every bit as bad as that sounds. (Yes, it's linked to cancer, but it also contributes to cellular insulin resistance, a precursor to diabetics.)

No wonder Phil Bruno felt so much older than his forty-seven years. His fat was, in effect, a giant toxic tumor that was poisoning the rest of his body. As his doctor warned him, he was at extreme risk for dying of diseases that normally afflict much older people, chiefly diabetes and cardiac arrest, but also stroke, cancer, and dementia. "Obesity has a lot of things in common with an accelerated-aging kind of situation," says Kirkland.

So aging makes us fat, and then our fat makes us age. And just as butter and lard get rancid with time, old fat causes us to age even more quickly. Kirkland believes that senescent cells buried in fat tissue may be the major culprits in the systemic inflammation that accompanies aging—and the older we get, the more senescent cells are lurking in our fat deposits. Another, equal problem is that our fat cells themselves become dysfunctional, and less able to do their job of, well, storing more fat. This leaves free fatty acids circulating around in our bloodstream, a dangerous situation called lipotoxicity—or, if you prefer, fat poisoning. Not good. And no wonder obesity has been linked to telomere shortening (which, in turn, creates more senescent cells).

This negative feedback loop only accelerates with age. Problem fat, though, overwhelmingly affects the middle-aged to older crowd, which is why the "metabolic syndrome"—a combination of obesity, insulin resistance, high blood pressure, and poor cholesterol—afflicts just 7 percent of people in their twenties, but nearly half of those in their sixties. Phil Bruno had the metabolic syndrome, and probably so does every heavy person you see at the mall. Another scary statistic: More than two-thirds of Baby Boomers are overweight or obese.

Indeed, there is growing evidence that our fat tissue itself might literally be shortening our lifespans. In a dramatic 2008 experiment, Nir Barzilai and his colleagues at Albert Einstein College of Medicine in the Bronx surgically removed the abdominal fat from a strain of obese laboratory rats, and found that the animals lived more than 20 percent longer than their still-chubby cousins. Their abdominal fat was, basically, killing them. As Barzilai puts it, "Not all fat is just fat."

Phil Bruno knew this all too well. Unfortunately for him, though, surgery wasn't really an option: In humans, says Barzilai, visceral fat cannot be safely removed, because it is so deeply entwined with our blood vessels and organs. Liposuction only removes "good" subcutaneous fat, which is why several recent studies have linked the procedure with what scientists call "poor health outcomes," and you and I call "death."

So in July 2004, roughly a month after he was diagnosed, Phil Bruno did the one thing that his doctor had *not* prescribed: He went to the gym.

You read that correctly: While Phil's doctor did suggest that he lose weight, he stopped short of actually recommending exercise as a way to do it. Incredibly, surveys show that only about half of diabetic patients are told by their doctors that they should exercise—likely because of the doctors' belief that their patients won't actually go to the gym, or that they won't stick with it, or that they won't end up losing much weight in the long run. There are studies that support each of these pessimistic points of view, too. But none of their authors had ever met Phil Bruno.

After a few weeks on his new medications, Phil had felt exactly as miserable as before, only now he was even more tired. Four of the meds carried drowsiness warnings, which meant that some days he felt like crawling under his desk and falling asleep by 11 a.m. He also knew that his medications were not treating the causes of his disease, only the symptoms. He felt trapped, hopeless, and depressed. "Something clicked in my head and said, *This is not gonna work*," he says. A devout Catholic, he turned to prayer in search of answers, as he wrote in an account of his struggle that he titled "The Jesus Christ Diet Plan":

> For me it all started with going to church and sitting slumped over in a pew having a heartfelt, tearful, prayer

> to Jesus. I just kept saying Jesus, Jesus, Jesus, over and over again. After being there for about an hour a fundamental question popped into my mind, I feel this clarity of thought was the first thing the Holy Spirit did to help me.
>
> The question was... <u>Do you want to live or die?</u>
>
> The answer... I wanted to LIVE!!!!!!!!!!!!!!

He realized that more than anything, he wanted to see his kids get married. The way things were going, it did not look like he would. For motivation, he turned to the Bible, but also to books he'd read by former football coach Tony Dungy and motivational guru Tony Robbins, who reinforced the message that his future need not be dictated by his past. It was what he needed to hear.

After getting his heart checked out (it was enlarged of course, from working overtime all those years, but his arteries were clean thanks to Grandma's olive oil), Phil walked into his local Gold's Gym one Saturday in July 2004. He looked around uncertainly before settling on the one thing that seemed doable for a 450-pound guy: the exercise bike. He hoisted himself aboard and managed to pedal for five minutes before he had to stop, wheezing and panting and feeling self-conscious—yet invisible at the same time. "Everyone looks at the fat guy in the gym," he says.

Yet he came back the next day, and the next. Soon he could stay on the bike for thirty minutes, leaving a bigger puddle of sweat on the floor each time. He visualized each drop of sweat as one more blob of fat exiting his body, one tiny step toward his goal.

On the way to his favorite bike, in those early weeks at Gold's Gym, he would walk past the glass-walled indoor-cycling studio. With the pounding music and the lithe bodies pumping away on the stationary bikes, it looked cool. It took him another week or two to work up the courage to show up for a Spinning class, and he made instinctively for a bike in the back corner. But he'd been spotted: A beautiful, fit blonde came over and confronted him. "I'm Beth," she said, smiling. "Let's help you get set up."

She put him in the front row. Her name was Beth Sanborn, and she looked every inch the Ironman triathlete she was. But she didn't care how much *he* weighed. He got through the forty-five-minute class, puffing and churning away on his bike. Phil soon became a regular in Beth's class, three times a week, and being Phil he soon got to know everyone in the room. "I had never seen anybody that big," Sanborn says. "He was the hardest-working person in my class—a man on a mission, he really was."

Phil gave himself Sundays off, but he kept coming in, day after day. He often rode until his shorts were bloody, because they don't make bike shorts (or bike seats) for people who weigh more than 350 pounds. "It wasn't pretty," he says. By September, he decided to take on an even bigger challenge: He would do a century, a hundred-mile charity ride to benefit multiple sclerosis, which his wife, Susan, had been diagnosed with. He hadn't been on a real bike in twenty years, but he dragged his old Trek out of the basement, dusted it off, and took it to the shop.

He made it all the way to Mile 63, on a slight uphill, when the road started to wobble and melt underneath him. He had pains in his legs and his chest, and, ominously, he had stopped sweating. The sag wagon was following him, and the event medical staff rushed to his aid, grabbing his arms to keep him from collapsing. "The thought actually went through my mind that if I die here on the road, at least I'm doing something to change my life," he says.

* * *

Without knowing it, Phil had kicked off a war for control of his body, with fat on one side, muscle on the other. He already knew that fat is stubborn stuff. "I'd been on every diet you can think of," he says.

Fat is bossy as well as stubborn. Much of the time, paradoxically, it's telling you to *eat more*, which is one reason why diets so often fail. Our fat wants to keep us fat. Although some fat tissue secretes leptin, which tells us to stop eating, very obese people become deaf or insensitive to leptin. So even as Phil Bruno was driving down the Manchester Road with a Quarter Pounder on his lap, his brain would be screaming that he was still hungry. While it would be hard enough for a "normal" heavy person to lose weight, for someone like Phil Bruno it would be next to impossible, says Mark Febbraio, a diabetes researcher at the Baker IDI Heart and Diabetes Institute in Melbourne, Australia.

"When you're talking about people who are over four hundred pounds, usually those individuals have a genetic defect in the signals coming from various parts of the body to the brain, that tell us to stop eating," Febbraio says. "And so they have insatiable appetites. Lifestyle can modify it to a certain extent, but if you're always hungry, eventually you'll start eating again."

Which makes the story of "A. B." all the more remarkable. Known to science only by his initials, A. B. was a twenty-seven-year-old Scotsman named Angus Barbieri who showed up at the hospital at Dundee, in northeastern Scotland, more than forty years ago. He weighed 456 pounds, which by the standard of the pre-obesity 1960s was downright freakish. With the encouragement of researchers, A. B. went on the simplest diet possible: He stopped eating.

He ingested no food at all, only vitamins and brewer's yeast, while the doctors monitored his health carefully. The weight came off, but more slowly than the scientists expected; after all, the guy was taking in almost zero calories, subsisting solely off his massive fat stores. He should have burned that fat right off himself. In the end, he managed to slim down to a very normal 180 pounds—an astounding achievement by any measure—but it took him 382 days to do it. Nor did he feel like he needed it. He did confess to "feeling a we bit weak," and he pooped only about every forty days, but other than that, he was fine–much to his doctor's surprise. Even more amazing was that the weight stayed off.

Phil Bruno wasn't about to quit eating. Rather, he made simple, sensible changes to his diet. Rather than starve himself, he started by cutting out fried foods, fast food, and sweet sodas, which had all been major components of his former food intake. He replaced them with things like grilled chicken and fish, while snacking on unsalted almonds rather than potato chips. A little common sense made a big difference. "The first fifty pounds just melted off," he says.

His initial goal was just to be able to weigh himself on his home scale, instead of at the grocery store. But he also loved food, and once in a while he'd eat an extra chicken breast at dinner, if he felt like it. Better that than a Quarter Pounder with fries. "When you're eating two Value Meals a day, with apple pie and a chocolate shake, any change is an improvement," he points out.

But as he kept exercising, Phil found that he not only lost weight, but also felt less hungry. In addition, his burning thirst was gone, and his long-suffering knees and hips felt better. He threw himself into his Spinning classes; eventually, he would become certified as an instructor, one of the most popular at that branch of Gold's Gym. "We saw quite a remarkable change," says Jim Wessely, a friend from the cycling class, who is head of emergency medicine at St. Luke's Hospital in St. Louis. "When he first came in he was this huge, morbidly obese guy who could barely spin for more than a few minutes; now he would really go at it."

As Phil puts it, "Quitting was not an option, because that meant death."

Ebullient and enthusiastic, he's a motivator, not a quitter; no surprise that he's in sales for a living. As he trained to finish his first century ride, in the spring of 2005, he organized his gym buddies into a cycling team called the Golden Flyers. Now they have more than 150 members who travel to charity rides all over the Midwest. He loved indoor cycling classes so much he trained to become an instructor, and was certified in 2008, four years after he walked into the gym as That Fat Guy. Three mornings a week, he now teaches the very classes he used to fear, leading a madcap, devoted crew of followers who show up at 5:30 a.m. to work themselves into a frenzy; Phil sets the tone with his favorite T-shirt, which reads, TRAMPLE THE WEAK, HURDLE THE DEAD.

He also belongs to a national cycling group for diabetics called Team Type 2, dedicated to helping people manage the illness through exercise rather than medication, which was how I found him. Phil holds the record for having lost the most weight. "He's something else," says Saul Zuckman, one of the leaders of Team Type 2.

If Phil was relentless, it's because his enemy was, too. In sedentary, inactive people—whether or not they are actually obese, like Phil—fat actually invades the muscles, slipping in between the muscle fibers like the marbling in fine Wagyu beef. Even worse, fat infiltrates the muscle cells themselves, in the form of lipid "droplets" that make the cells sluggish and may even contribute to insulin resistance, says Dr. Gerald Shulman, a prominent diabetes researcher at Yale.

According to Shulman, these "pools" of fat, in both liver and muscle, block a key step in the conversion of glucose, thus leading to the insulin resistance that is a prerequisite for diabetes. This also explains why many sedentary normal-weight people are still at risk for the disease. "It's not how much fat we have, it's how it's distributed," Shulman says. "When the fat builds up where it doesn't belong, in the muscle and liver cells, that's what leads to Type 2 diabetes."

By exercising so intensely, Phil was vacuuming up those pools of excess fat. As a result, his insulin resistance and his diabetes seemed to be easing. Instead of floating around his body wreaking havoc, his excess sugar was being incinerated in the furnace of his muscles. He was still overweight—his weight fluctuated in the high two hundreds—but he was now in a completely different metabolic state. Recent data shows that while being obese is generally a serious health risk factor, the small category of "fit and fat" people have much less to worry about.

A year after he was diagnosed, he went back to Dr. Livingston for some routine tests. The doctor was astonished: Bruno's insulin resistance was gone, and his blood values were almost back to normal. His A1C, which had been 16, was now down to a more normal 5.5. He'd never seen anyone do that. Reluctantly, he took Bruno off all his medications. He had controlled his diabetes, which once threatened his life, by changing his diet and embracing exercise. It had been almost like a second job. "I'm a financial adviser with Wells Fargo," he says, "but most people think I'm a Spinning instructor."

Yet Phil knew he was far from "fixed." Because of his individual makeup, he'd been primed for weight gain for his entire life—even as a kid, he wore husky-size pants. He had to fight a constant, escalating battle against his morphological fate. He stayed dedicated to the gym, going to Spinning classes four and five days a week; on Sundays, he organized a regular group ride, and he mustered Gold's Gym fund-raising teams for charity rides like the Tour de Cure (for diabetes), and the MS 150.

In the spring of 2015, Phil Bruno realized another lifelong dream when he bought himself a brand new Harley-Davidson. With his son, Phil Jr., he flew to Seattle and together they rode their bikes across Washington, Oregon, and into Montana. They toured Yellowstone National Park before continuing east to Sturgis, SD, in time to join a million other motorcycle enthusiasts at the massive annual Sturgis rally. Then they turned south and made their way down the Missouri River to St. Louis, for a total of 2,500 miles.

Phil seemed happier than he'd ever been; had grown a new beard and sported motorcycle leathers, a counterpoint to the Lycra he still rocked in his 6 am spin classes. His wife, Susan, often accompanied him to class, bravely fighting the effects of her MS, even as the disease slowly worsened. On the way out to the parking lot of the gym or when they went to restaurants, Phil always made sure he was at her side, discreetly holding her by the elbow to steady her.

On Friday morning, August Phil taught a spinning class; that evening, there was a luau party at the Harley dealership (where his son worked). He was videoed hula-dancing. On Saturday, though, he woke up feeling unwell. The feeling worsened through the day, and after Sunday Mass he was taken by ambulance to St. Luke's hospital, where he died of a massively torn aorta. There was nothing that could be done. He was 58 years old.

"He was selling his lifestyle to everybody," says his son, Phil Jr. "But he was on a roller-coaster ride the entire way."

Chapter 10

POLE VAULTING INTO ETERNITY

"To be seventy years young is sometimes far more cheerful than to be forty years old."

—Oliver Wendell Holmes

On a cool, overcast Cleveland summer morning, I stood in the infield of a small college football stadium, watching some of the nation's top athletes battle each other in an important track-and-field meet. The atmosphere was intense, and in between events I gravitated to the infield, where I met three lanky sprinters preparing for their heats. Their names were Ron Gray, Don Leis, and Bernard Ritter, and they still wore their warm-up suits against the early-morning chill. As they stretched and limbered up, they coolly assessed their competition while, like male athletes everywhere, saving a glance or two for some of the more attractive female athletes.

Ron, Don, and Bernie had all distinguished themselves in their events. Don had set national records in the triple jump, while Ron was one of the top sprinters in the country, and Bernie was competitive at the state and regional level in his native South Carolina. They had each trained for months in the hope of medaling in this national-level meet; their next stop would be world championships in Brazil later in the year. Don came from the track-and-field hotbed of Pasadena, while Ron hailed from Denver, where he had played running back for the University of Colorado football team.

Sixty years ago.

The event was the 2013 National Senior Games, a kind of biannual Olympics for older athletes ("senior" being defined as anyone over fifty). It featured the traditional track-and-field events and swimming, but also triathlon, basketball, volleyball, badminton, and Ping-Pong, to name a few. And of course shuffleboard. The competition in pickleball, a kind of paddle tennis that is becoming hugely popular in retirement communities, was said to be especially fierce. To get here, competitors had to qualify at state and local Senior Games, which meant that Cleveland was playing host to the crème de la crème of older athletes.

I focused on track and field, because its results are quantifiable—there is a finish line to cross or a crossbar to clear—and because it takes no special talent. Anybody can run and jump. Which is why I found myself in the middle of this stadium at nine in the morning, surrounded by much older people wearing Lycra, track spikes, and very serious expressions. Also hearing aids. It was as intense as any college track meet, just a little more wrinkly.

I had come here in search of people who were "aging successfully," in the patronizing lingo of the geriatricians, and I found way more than I bargained for. Over the course of the weekend, I watched a nearly ninety-year-old woman heave an Olympic javelin far enough to clear a double-wide, the long way. The best high jumper in the seventy-through-seventy-four age group cleared a mark that would have won him a silver medal in the 1896 Athens Olympics, and I watched a ninety-two-year-old man trot down a narrow runway and attempt to pole vault a bar set as high as my head.

This terrified me. Was he nuts? Who had allowed this to happen? The whole morning had stoked my anxiety, as I stood there watching people my parents' age sprint and jump so hard they almost puked. I couldn't even remember the last time I'd run a hard quarter mile. One thing these athletes all had in common was that they were breaking the rules—not the rules of track-and-field competition, but the unwritten rules governing acceptable conduct for so-called old people. Everybody claps for the grandmas shuffling through the local 5K. But when Grandma can smoke her middle-aged kids in the hundred-meter dash, things can get awkward. It just seems wrong, if not downright dangerous. (This seems an appropriate place to add: By purchasing this book, you have agreed to absolve the author of any liability for whatever happens to you while "trying this at home.")

But these were not people who cared what their kids think. There in the infield, Ron, Don, and Bernie talked about training and diet like seasoned competitors, which they were. This was Don's eighth meet this year, and it was only July. Today he was a bit under the weather, thanks to a black eye he had acquired while "chasing my grandkid around the playground." Ron had the leathery complexion of a man who had spent a bit too much time in the Rocky Mountain sun, which he had. But the rest of him still looked like the University of Colorado football recruit he had been in the early 1950s, with his barrel chest and powerful forearms atop a pair of tapered, spindly legs. He had run track in college, in addition to four seasons as a running back for the Buffs. "Then nothing, for fifty-five years," he laughed. Although by "nothing," he actually meant hiking up Aspen's Highlands Bowl and skiing down its forty-five-degree steeps, on a pretty much weekly basis.

Four years ago, a college classmate suggested Ron enter a Masters track meet. He got hooked on the competition, which reminded him of his days in the auction business. And he loved the training. Now he works out three days a week on a local track, all by himself, plus two days in the gym with a trainer. "It makes you feel good, the heartbeat and the butterflies in your stomach, and all that," he'd tell me later. "And beating the other guy, of course."

One thing that struck me was that Ron and Don and Bernie all approached aging with the same discipline that they put into their training, as though aging itself were a kind of athletic event. Ron stays young, he said, by avoiding the "inflammables," by which he seemed to mean foods containing dairy, wheat, and sugar. He'd been experimenting with this new anti-"inflammable" diet for the past six months, he said, and it seemed to be working. "I woke up one morning last year, and nothing hurt," he deadpanned. "I thought I was dead. Now I wake up and nothing hurts because nothing hurts!"

And come to think of it, he *did* look a little like Rodney Dangerfield. He also employed the services of a "concierge physician," a high-end doctor who was basically on retainer. He could go and see him for anything, anytime, and stay as long as he wanted. Education, money, and access to medical care all correlate strongly with longevity, Jay Olshansky and other scholars have found. Ron, Don, and Bernie had all three.

The trio planned to enter the 4x100 relay as a team, but they needed a fourth. I was secretly relieved to be four decades too young. Their real enemy, though, was not the other runners, but injury. And indeed, as we stood there chatting in the infield, the boys and I watched a seventy-five-year-old woman stumble during a hundred-meter heat, hit her head on the track, and get hauled off in an ambulance, still unconscious. In the very next race, a gentleman about the same age pulled his hamstring at the start and fell into the gutter, writhing in agony. "He pulled it *bad*," said Ron with a grimace as the EMTs came trotting back out with their stretcher.

Ron knew better. He warmed up carefully before every workout and every race, using a specific routine designed to prepare his eighty-one-year-old muscles and the ungodly stress of sprinting. As his start time approached, Ron went through his warm-up routine, as always, in his red USA jersey and Lycra compression shorts, as the other age groups went off, one by one. Then it came time for the men's eighty-through-eighty-four age group, and he sauntered into position as his name was called. Lane 6.

"Take your marks," said the starter.

Ron crouched down, placed his fingers on the nubbly track surface, and set his feet in the starting blocks: front, then back. Two of the runners didn't bother using the blocks; at eighty, it can be a bit problematic to spring out of a crouch. So they stood on the line, which was easier on their hamstrings but also removed them from serious contention. That left three runners in the mix: Ron, a lean African American man named Alex Johnson who had crushed him in qualifying, and another runner named John Hurd, from Florida, who had also beaten Ron in the past.

"Set," the starter said, and the runners froze. Down at the finish, a small knot of spectators—mostly wives and middle-aged kids of competitors—lounged in the stands, looking a bit bored.

BANG!

Ron burst out of the blocks, a move he practiced several times a week on the track near his home. Within two strides, he was fully upright, running as fast as he could, his arms slicing back and forth, his palms like knives to cut through the air while his legs whipped around like the blades of an eggbeater. And by ten yards out, he was leading, a stride or two ahead of both Hurd and Johnson. But Johnson came past him at seventy-five meters, devouring the track with his huge strides. Ron hung on for second place, clocking a time of 16.75 seconds, nearly a second faster than he'd run at Masters Nationals three weeks previously. He was happy, as he stood in the infield and tried to catch his breath.

"I can't wait for the next one," he said, still panting.

Athletes understand aging better than almost anybody, because they feel its effects sooner than the rest of us. A pro football player will be considering retirement at thirty; LeBron James is getting on in years at thirty-one. Endurance sports are more forgiving, but not much. Meb Keflezighi won the Boston Marathon at age thirty-nine, which was hailed as an amazing feat. The oldest rider in the 2014 Tour de France was forty-three years old. Jamie Moyer was fifty when he became the oldest pitcher ever to start a Major League Baseball game. He pitched two innings.

For professional athletes, remaining healthy with age is vital to their livelihoods, which is why former Yankees star Alex Rodriguez and frequented South Florida "anti-aging" clinics in search of human growth hormone and other chemical magic that might help prolong his career. For Masters athletes, though, the dynamic is reversed. They are no longer victims of age but combatants, battling with it as the decades slip by.

"To become an athlete at age 47, or 50—or 90, I'm sure—is merely a way of saying '*Wait!*'" wrote the late John Jerome in *Staying With It*, his wonderful memoir about taking up competitive swimming late in life. "It is a way of grabbing time by the lapels, of saying stop, wait a minute, let me understand what is happening here. Maybe the point isn't to fight age off but to let it come on, to get inside it, to find out just what it is."

Few Senior Games competitors had gotten as far inside aging as Howard Booth, whom I met later that afternoon in the "pits," a grassy area where the jumping events were held. The men's long-jump competition was in full cry, with gray-haired guys charging full speed down a narrow runway and launching themselves into a pit of sand. My knees throbbed just watching them. Booth distinguished himself with a very special jumping style: When he landed, he'd turn a quick little somersault and pop right back up onto his feet. This made the spectators and even the judges laugh every time.

Booth was not only a former college gymnast—hence the tumbling—but also a professor of biology at Michigan State University, with a keen personal and professional interest in both athletics and aging. Compact yet muscular, with white hair and a clipped beard, he wore a skintight bodysuit that showed he was in enviably good shape for any age. His specialty is actually the pole vault, which is a bit odd considering he is not particularly tall, but what he lacks in height he makes up for with passion.

He had pole-vaulted in college, but quit to pursue his graduate studies and his research. About ten years ago, a friend told him about the Senior Games, and for fun he checked out the pole-vault records for his age group. They seemed well within his reach, so he decided to pick it up again. He built a pit in his backyard, making uprights out of scrap lumber and a landing pit of trash bags filled with leaves. A maple sapling served as his pole, and another as the crossbar, and boom, he was a pole vaulter again. He's since upgraded to a more professional-quality setup, and now his backyard pit attracts vaulters of all ages on Sunday mornings. He medals in national events on a regular basis.

"You can wake up with sore muscles and ask yourself, *Why am I doing this?*" he said. Answer: "Because flying up in the air is really, really fun. Mentally, we're kids playing."

There might be something to the "mentally" part. It recalls the famous experiment done by Harvard psychologist Ellen Langer, who recruited eight older men to spend a week together in a house decorated, in every detail, in the style of their 1950s heyday. Even the magazines and books were from that decade. The men were then instructed to imagine themselves as they had been in 1959, when they were in their prime (to aid this, all mirrors had been removed). They discussed 1950s sports and news as if they were in the present, and so on. She told them to "inhabit" their former selves.

By the end of the week, the men had been miraculously rejuvenated, performing far better on Blast-like tests of grip strength and such, and even breaking out into a spontaneous game of touch football. They were miraculously rejuvenated. "It almost seemed like Lourdes," Langer said later.

* * *

Hippocrates believed that exercise was medicine, and so did the physicians of ancient China. It went out of fashion in the early twentieth century, though, and was actually believed to be dangerous—coincidentally, just as heart disease was emerging as a leading cause of death. During the first half of the twentieth century, doctors typically prescribed bed rest for their patients with heart trouble. Oops.

That changed in the 1960s, when the massive Framingham Study found that people who exercised regularly were far less likely to suffer heart attacks than those who did not. Those who smoked, on the other hand, were at greater risk. Since then, a tidal wave of exercise data has all pointed in the same direction. A recent analysis of statistics covering more than 650,000 individuals showed that people who kept to a normal weight and exercised moderate-ishly, the equivalent of a brisk walk for an hour or so per day, lived an average of *seven years* longer than the non-exercisers. There is a raging debate over whether or not more intense or long-lasting exercise confers proportionally greater benefits, but one study of Tour de France veterans found that they, too, lived about seven years longer than their peers. So did Olympic medalists, by three years, according to a study of more than 15,000 athletes from 1896 through 2010.

That may be due to all the wine, at least in the cyclists' case, but it's more likely because of the fact that exercise itself is literally like medicine, as a growing body of evidence is beginning to show. In a detailed and revealing comparison, the Stanford scientist John Ioannidis paired more than three hundred randomized clinical drug trials with the results of fifty-seven studies of exercise, and found that in nearly every case, exercise proved just as effective as the medications, and sometimes better, at staving off death from heart disease, stroke, and diabetes.

"If you could put the benefits of exercise in a pill, it would be an astonishing pill," says Simon Melov, a researcher at the Buck Institute who has studied exercise extensively. "The data is now coming out on the effects of chronic exercise, and it is astonishing in terms of its ability to prevent all sorts of age-related disease, everything from cancer through to neurodegenerative disease to heart disease, even arthritis. All of these things have vastly lowered risk in people who exercise regularly—and if that was in a pill, it would be in*sane*."

As a biologist, Howard Booth already knew this. After college, he'd kept active by running and biking, so he was still in relatively good shape, but he found himself getting bored. He took up pole vaulting at age sixty in part because he knew he would need a new challenge when he retired from teaching. "I reflect on my father's generation, where the idea of retirement was that you've worked really hard all these years, and now you deserve to do nothing," he told me. "Not something else, but *nothing*."

That wasn't for him. Neither was golf, once the sole socially acceptable pastime for men in their sixties. "Just a step up from watching paint dry," he scoffed. Though he's not fully retired from teaching (and coaching) at Eastern Michigan, training and competing in pole vault gave him another goal, or as the Okinawans call it, *ikigai*, a sense of purpose. The stakes are low—at the Senior Games, he stood to win a cheap alloy medal, at best—but at the same time, they couldn't be higher. "Even if it's a two-dollar ribbon, and you spent thousands in travel and overnights," he said, "it makes no sense—but it makes huge psychological sense. You're not on the sidelines, you're really participating."

Something similar was taking place biologically, inside him. He was *participating.* As he perused the results and record books, he grew curious: Just how good was it possible to get? Could an older athlete ever equal his prime? Could he or she come close?

Booth had his students survey his peers on the senior track-and-field circuit, and found that dedicated athletes in their sixties could perform about 80 percent as well as they could in their prime. That is, if they could vault fifteen feet in college, they were hitting twelve feet in their sixties. National records also bear this out: In the men's hundred-meter dash, for example, the best time recorded by an American sixty-through-sixty-four-year-old in 2012 was an astonishing 11.83, barely 2 seconds slower than Justin Gatlin's Olympic gold medal–winning 9.79. Among runners a decade older, seventy through seventy-four, the fastest man takes just a second longer (12.90), which is still pretty fast.

Top-level performance drops off pretty drastically after age seventy-five, but the really interesting fact that Booth uncovered was that his "controls"—that is, average sedentary adults of the same age—had retained only 22 percent of their physical capacity. This told Booth that the stakes were much higher than simply breaking records and winning medals—both of which he did with relative ease. It was more about staying in the game, about not giving up. "Most people just aren't basically healthy at this age," he said. "And the idea that well, it's natural and you're just an old man now—that *isn't* natural! That is the default. That's where we get to by not challenging ourselves. Exercise is a continuum: The more you do, the less you're going to lose."

Booth had finished his events for the day, and we were sitting in folding chairs, relaxing in the sun and talking about the science of sport. His wife, Luanne, sat nearby as he talked, nodding. "We can show it in muscle protein production; in nerve junctions, that tend to fade away; and the rate at which muscle fibers return," he continued. "All of these basically respond to increased exercise, by slowing the rate at which they would naturally decline. And the greater intensity you can put into that package, the slower you will lose performance."

He had chosen pole vaulting because it requires not only basic fitness, but physical skill and fine coordination, which—as The Blast has shown—decline even faster than aerobic capacity. The fast-twitch muscle fibers used in jumping and sprinting tend to disappear earlier than the slow-twitch fibers used by endurance athletes. "The more you use them, for fine detail—the precision of a fine overhand in tennis, or a jump shot—the fast-twitch fibers in particular will have greater numbers of motor units [a combination of muscle fiber and the nerve that triggers it] in the areas you have used and worked on," he said. "And if you quit doing those things, they decline."

While I was working on this book, nearly everyone I told about it wanted to know the same thing: "So, what's the secret to aging?"

So far, the "secret" seems to be: *Use It or Lose It.*

Which sounds simple, even simplistic. But it kept coming up, almost like a mantra, not only in conversation but in high-level research: It applies to your cardiovascular system, your muscles, your sex life, and your brain. Howard Booth had it all figured out.

By contrast, *not using it* can have dire consequences. Even retiring from working—the capstone to the American Dream—can be dangerous to your health. A paper published by the National Bureau of Economic Research, a prestigious private think tank, found that "complete retirement leads to a 5–16 percent increase in difficulties associated with mobility and daily activities, a 5–6 percent increase in illness conditions, and 6–9 percent decline in mental health," over the next six years. Although early retirement has been found to decrease mortality risk, at least in Europe, newly retired people often report a loss of their sense of purpose—the Okinawans' *ikigai*, again—which can be hard to replace.

As everyone since Brown-Séquard has noted, physical parameters like strength and VO_2 max tend to move in one direction with age: downward. But it's not the same for everyone. A recent study of aged Scandinavian cross-country skiers found that the older athletes had preserved much of their aerobic capacity, relative to their youthful selves; and they were far ahead of the age-matched control group, a bunch of sedentary older guys who lived in Indiana.

Which seems like the ultimate unfair comparison—Nordic ski gods versus Midwestern couch potatoes—but who would you rather be? The skiers had done a better job preserving their ability to pump blood efficiently, the elasticity of their arteries, the suppleness of their lungs. Biologically, they were simply younger. On a practical level, this meant that they had an easier time walking around, climbing stairs, and as Howard Booth put it, *participating* in life. They'd never stopped using it, so they didn't lose it.

If you look at older athletes' muscles and bones, the contrast with their sedentary peers becomes even more dramatic. One of the hallmarks of middle age—and one of the first things I noticed—is that it becomes much more difficult to gain and keep muscle. We begin to lose muscle mass gradually at around age forty, and as time goes on we lose it more rapidly: Between fifty and seventy, we say good-bye to about 15 percent of our lean muscle per decade. After that, it jumps to 30 percent per decade. "You could make the case that aging starts in muscle," says Nathan LeBrasseur, a researcher at the Mayo Clinic who studies muscle.

But even as we're losing muscle in middle age, we don't lose weight overall (duh). That means our muscle is gradually, insidiously being replaced by fat. More fat and less muscle means your metabolic "engine" runs at a much slower rate; you have less muscle, which means you have fewer mitochondria, which means your body is less efficient at burning the sugar out of your bloodstream. Not coincidentally, most new cases of diabetes appear in people in their mid-forties and older. And this loss of muscle may be why our cholesterol levels tend to surge as we get older.

The reasons for this are not fully understood. Lower testosterone levels have been singled out as the culprit in middle-aged muscle loss, but hormonal changes aren't the only guilty party. Older people are up against an even more powerful enemy, something in their own bodies. Experiments involving parabiosis have revealed that in older mice (and presumably also in older humans), muscle stem cells or "satellite cells" have a harder time activating in response to injury or stress, because of something that circulates in old blood—or is missing from old blood. As if that weren't bad enough, our bodies also produce a hormone called myostatin, whose job is actually to *slow* muscle growth—to keep us from growing too big and thus requiring too much food, according to LeBrasseur. (Thanks, evolution.)

If this muscle loss continues and accelerates, it puts us at risk for a condition called sarcopenia, or what Shakespeare—keen observer of old age—called the "shrunk shank," where our limbs basically waste away, putting us at risk for frailty. Loss of muscle due to age is why my surviving dog, Lizzy, now has difficulty jumping up onto the bed or into the car, where she used to clear five-foot fences on the fly. She's lost her spring. Her haunches, once firm, have gone soft. It's not just about pole vaulting or fence jumping, either. People (and dogs) with sarcopenia are at greater risk for falling, and in the frail, a simple fall can snowball into a fatal event; this is why muscle wasting is the second-leading cause of institutionalization of the elderly, after Alzheimer's.

Nobody knows quite what causes sarcopenia; even its exact definition is controversial among scientists. The cure, too, is a subject of debate. For people like Dr. Life or Suzanne Somers, the answer is easy: Shoot up with testosterone and growth hormone. By this point in the book, you should realize what a bad idea this is. And anyway, while replacing testosterone does increase muscle size, it does not always improve muscle *quality*, at least not without exercise.

Half a dozen pharmaceutical companies are developing drugs to fight sarcopenia in the elderly by promoting muscle growth; the drugs are only in early-stage trials, but some of them have already appeared on the black market for athletes and bodybuilders. But there is another, simpler treatment for sarcopenia that science has largely ignored, until recently: staying active. Adults who have exercised for most of their lives keep muscle for longer, as the illustration on the next page shows rather dramatically.

These drawings are based on MRI images of the upper legs of four different men, similar to those taken in The Blast. Each is a cross-section of someone's upper thigh. The one on the upper left belongs to a typical fittish forty-year-old, and you can see how it's

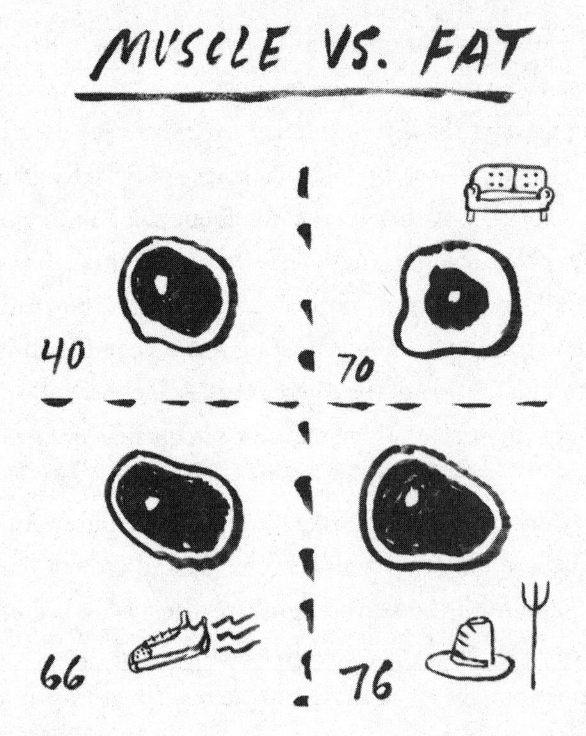

mostly muscle, with a small ring of subcutaneous fat on the outside. The one at upper right is a typical sedentary seventy-year-old American man, with the classic signs of sarcopenia: Note how it is nearly all fat, like a slice of pancetta from a very well-fed pig. But the fat has also completely infiltrated his muscle, making it look "marbled" and rendering it weak.

The bottom two images are just as striking: On the left, a sixty-six-year-old triathlete—looking pretty much the same as the fit forty-year-old. (According to Nathan LeBrasseur, the differences between them would only be visible through a microscope.) But the one on the right belongs to a *seventy*-six-year-old man who has never done a triathlon or any other sort of running in his life. He's an English farmer, whose job required him to stay on his feet most of his life, moving around every day. He's got about the same "muscle age" as the forty-year-old, because his lifestyle is the closest to that of our evolutionary ancestors.

By pole vaulting and jumping and sprinting into old age, Howard Booth and his fellow senior athletes are not only defying time and gravity, but imitating, in a way, the kinds of things our hunter-gatherer ancestors had to do. Staving off sarcopenia is just a side benefit. They'll be able to chase their grandkids around on the playground, or even just walk around, say, Paris while on vacation. The sedentary man won't. *Use it or lose it.*

Use It or Lose It even applies to laboratory mice. Traditionally, mice are kept in small plastic cages slightly bigger than shoe boxes, with nearly unlimited access to food but no opportunity to exercise. They live alone, too, because males in captivity have a tendency to fight. About a decade ago, a lab assistant to Tom Kirkwood named Sandy Keith did a small, unpublished experiment where he simply gave his mice bigger cages and more toys to play with, simple things like cardboard toilet-paper tubes. He also played with the mice every day, keeping them socially engaged. One of these "free-range" mice, a male named Charlie, lived an astounding 1,551 days, or four years and three months—which is six months longer than even the longest-lived caloric-restriction mouse. He had a lot more fun, too.

To this day, Charlie remains one of the longest-lived mice ever—simply because he was given the opportunity to "use it." The problem is that most older people aren't expected or encouraged to do much of anything, and it often hurts when they do, so they don't. "For some generations, exercise has a stigma. You say 'exercise,' it turns people off," says LeBrasseur, whose research center is connected to a senior living community. "The most striking thing to me is how people are building their world around the La-Z-Boy. They have their medication around them, and the TV, and their food. They've engineered the physical activity out of life."

So it's not even "exercise," in the sense of slogging away on the treadmill at the gym, but something closer to "just moving around," like the English farmer. Recent studies have pinpointed the mere act of sitting, itself, as a potent risk factor for death. An analysis published in the *Lancet* in 2013 found that inactivity was responsible for more than 5.3 million premature deaths each year worldwide, from causes ranging from heart disease to colon cancer. The authors concluded that eliminating inactivity—perhaps by shackling people to treadmills? Banning televisions?—would reduce the instance of those diseases, as well as Type 2 diabetes and breast cancer, by 6 to 10 percent. Not only that, they concluded it would increase worldwide life expectancy, for the entire human race, by close to nine months.

Sitting is the new smoking, some scientists believe: a bad habit that leads inevitably to disease. Now go take a walk. Just be sure to hold your breath as you walk past the people who are actually smoking on the sidewalk outside the building.

At any rate, it's clear that moving muscle does something far more profound than simply burning up calories. Older people who exercise tend to avoid the hyper-cholesterolization that afflicts their sedentary peers, for starters. And it goes much deeper: LeBrasseur and his colleagues recently finished a novel experiment that dramatically illustrates the metabolic power of exercise. In the lab, LeBrasseur fed mice a special diet designed specifically to mimic the nutritional content of a fast-food meal: Big Mac, fries, and a Coke. The mice had been genetically modified so that any senescent cells would bind to a special fluorescent marker that would make them glow in the dark. After a few months on the all-fast-food diet, the mice lit up bright green, because they were filled with many more senescent cells than the mice who had been fed a normal diet. But the Big Mac mice that had also exercised had many fewer senescent cells. The exercise had negated the toxic effects of the #1 Combo Meal—either by zapping the resultant senescent cells, or by preventing their formation in the first place.

"It really highlights the power of exercise," he says. "You're pouring this toxic substance into your body, but as long as you're exercising, it's not going to be as bad for you."

So it's okay to go to McDonald's, as long as you jog there. (Or better, jog back.) But what scientists are discovering is that it's not just that the jogging is cleaning the Special Sauce out of your arteries; it's that your muscles are somehow communicating with other organs of your body to optimize their function. This we know thanks to innovative experiments in the 1990s with athletes who had been paralyzed due to spinal injuries. When their muscles were stimulated, in a way that mimicked exercise, researchers found that their livers "knew" to send a shot of fuel directly to their muscles. In the past, it was thought that this communication happened via the nervous system and the brain, but the spinal patients got the same burst of fuel; they even experienced a "runner's high." How could that be?

In 2003, biologists Mark Febbraio and Bente Pedersen found that just as fat "talks" to the rest of your body—usually saying terrible things—so does muscle. "We discovered that muscle, when you contract it, is actually an endocrine organ, and it can release factors that can talk to other tissues," says Febbraio. "So when a muscle contracts, it's not just an organ of locomotion."

The primary signaling factor they identified was a surprising one: our old friend IL-6, the well-known inflammatory cytokine that's normally associated with bad things like inflammation and early death. Exercise generates huge amounts of IL-6, they found, but in that context it actually has beneficial effects, such as signaling the liver to start converting fat to fuel. "When we made this discovery, people really didn't believe us, because IL-6 was considered a bad actor in many diseases," he says. "But the thing is, in exercise it's *anti*-inflammatory, actually."

The difference had to do with time. Obese and elderly people tend to have constantly elevated levels of IL-6, a sign of chronic inflammation. Normal-weight and younger patients have lower levels—but when they exercised, their IL-6 levels would spike to a very high level, and then dissipate over a few hours. These short bursts of IL-6 were in effect sending messages to other organs, like the liver and the gut, telling them to switch to "exercise" mode.

Since then, dozens more of these muscle-specific messengers, called myokines, have been identified. Febbraio, a former professional triathlete who describes himself as an "exercise junkie," believes there are hundreds more yet to be discovered, and that they are largely responsible for the myriad and complex beneficial effects of exercise. Some of them even act on the brain, triggering the release of BDNF, brain-derived neurotrophic factor, which heals and protects neurons.

In a sense, exercise helps the body clean house. Intense activity triggers a cellular cleaning process called autophagy, from the Greek words for "self-eating." Autophagy is crucial to our cells' survival. Without it, our cells would quickly fill up with garbage and become dysfunctional, just like your house would if you quit taking out the trash. "Exercise is an incredibly effective mechanism to drive protein turnover, kind of flushing out the old proteins," says LeBrasseur. It helps our cells clean house, so they can function better for longer.

Other myokines appear to work on bone, on the pancreas (which secretes insulin), on the immune system, and on muscle itself, promoting growth and healing. "It seems that muscle is the organ that counteracts fat," says Pedersen. Literally: One newly discovered myokine even attempts to convert fat to an energy-burning system, like muscle. In 2012, a Harvard-based team identified a hormone called irisin, secreted by muscle during exercise, which tricks plain old white fat, which is most of our fat, into acting like "brown" fat, a far more rare form of fat tissue that is dense with mitochondria and actually burns energy. Bruce Spiegelman, the Harvard scientist who discovered irisin, is now looking for a drug compound that might trigger its release, independent of exercise.

But Febbraio cautions that "exercise in a pill" is not in the cards: "It'll never happen," he says firmly, "because the benefits of exercise are a multifactorial thing. You could never design a drug that would replace exercise." Just ask Phil Bruno. For him, in fact, exercise wound up replacing the drugs.

Don and Ron and Bernie and Howard Booth all had one thing in common: They each looked, and acted, much younger than the age on their driver's license. Yet until recently, mainstream science has insisted that exercise doesn't really affect the aging process itself; it merely extends healthspan and improves function. In mouse studies (which researchers so love), it only appeared to increase *average* lifespan, not maximum; that meant it wasn't actually slowing aging, even if it did help certain individuals live longer than they otherwise would have.

But some new research is hinting that exercise may have a more profound effect on the aging process than scientists were previously willing to believe. In 2007, Simon Melov was part of a team that found that exercise actually seemed to reverse the effects of aging in a group of older Canadians. The study looked at two groups of people, one older and one younger. The researchers took muscle biopsies from each group, a painful procedure involving a rather large and long needle, and analyzed the "gene expression" patterns in the tissues—which genes were turned on and off. (Over time, different genes are activated in different cells of our bodies, a process called "epigenetic" change.)

They then placed half of each group on a strict but not-too-demanding resistance-exercise program for six months. At the end of the six months, they took more biopsies and found that the older subjects' muscles had reverted to a "younger" state—that is, they had had many of the same genes activated as their younger study mates. "We showed you could essentially reverse the gene expression signature of aging with exercise," says Melov.

In short, exercise had switched the "young" genes on, and the "old" genes off. Most of those genes had to do with the function of mitochondria, which you surely remember from high school biology class: They're the little energy plants inside our cells. I like to think of them as tiny little cellular turbines, but their history is what's really interesting. Mitochondria originally evolved as entirely separate organisms—parasites, really—back in the primordial soup days. Back then, most of the bacteria that made up life on earth were anaerobic: They survived without oxygen. But as the atmosphere slowly became more oxygenated, in this first great episode of global climate change, the anaerobic critters began to die off—unless they had been invaded by the tiny, oxygen-burning parasites that we now know as mitochondria. Now nearly all life depends on oxygen, and mitochondria are found in nearly every living cell. Yet because mitochondria are so ancient and primitive, they still maintain their own small, separate genome, distinct from ours, with just thirteen genes that are both highly important and highly fragile.

Melov's coauthor, a researcher and physician named Mark Tarnopolsky at Ontario's McMaster University, was fascinated by mitochondria, so he picked up where the 2007 study left off. Tarnopolsky had devoted his career to studying rare mitochondrial diseases in children and adults, and he noticed that many of his patients suffered from effects similar to a kind of accelerated aging: They developed premature gray hair, or went blind in their twenties, or lost muscle strength in their forties. Which makes sense, because mitochondrial dysfunction is believed to be one of the primary drivers of aging. "What is it about the aging process which unmasks so many of our disorders?" he asks.

Tarnopolsky also understood mitochondria on a more functional level, because he was a top-level amateur athlete himself, nationally ranked in cross-country skiing and trail running. As he neared forty, he started wondering about the effects of age on his own mitochondria—and whether exercise would keep those consequences at bay. One theory of aging says that over time, our mitochondria accumulate mutations to their own rather fragile DNA, which causes them to crap out, one by one. As we lose mitochondria with age, we eventually run out of energy, as Luigi Ferrucci observed. Making matters worse is the fact that our mitochondria are the site of some of the most intense chemical reactions in the body, which produce toxic free radicals and other damaging molecules. When enough mitochondria stop functioning, so does the cell—be it a muscle cell, a brain cell, or some other kind of cell.

Tarnopolsky wanted to try a study of mitochondrial aging in mice, but found himself unable to get funding—because, he thinks, the scientific establishment is biased against exercise studies. "Unfortunately, when you want to study exercise, you are already behind the eight ball," he says. "People say exercise is 'dirty,' because it has so many pathways. We get many of our papers rejected because people say, 'You haven't shown the exact mechanism of the pathway of exercise.'"

With no alternative, Tarnopolsky decided to fund the study himself, using more than $100,000 in profits from his medical clinic to pay for a small number of very expensive genetically modified mice. The mice had been programmed to undergo mutations to their mitochondrial DNA at a much greater rate than normal, which caused their mitochondria to conk out more quickly—which, in turn, made them age more rapidly. He then placed some of those mice on a regular treadmill-exercise program, just forty-five minutes three times a week, while letting others stay sedentary in their cages.

The results were dramatic. As expected, the sedentary mice were prematurely aged: gray, emaciated, and feeble. But the mice that had exercised were still strong, active, and sported lustrous black fur; they literally walked all over their sedentary cousins, despite the fact that they had the same broken mitochondrial DNA. The effects were far more than superficial, too: Autopsies showed that the exercise mice had stronger hearts (of course), but also healthier livers, brains, and even more robust gonads (yay!) than the inactive mice. Exercise had somehow repaired their mitochondrial DNA—in short, it had reversed their aging. Tarnopolsky suspects that in exercise, the mitochondria send some kind of signaling molecules that do the repair work in other organs, not just muscle. He was determined to figure out what those molecules were, because they might lead to drugs that could help his rare-disease patients. For the rest of us, though, exercise *is* the drug.

"It's very simple," he says. "Get off the couch."

I hung around the Senior Games for three days, watching these incredible older athletes run, jump, and throw things. At a certain point, I got used to it, almost numb to the sight of a seventy-year-old lady in a tennis outfit executing a perfect Fosbury Flop. But on the second afternoon, the entire stadium stopped to watch one extraordinary race.

It was the women's eight-hundred-meter, two laps around the track, and right from the gun one runner stood out from the rest. She was tall, with long silver hair, and her long legs ate up the distance in long, graceful strides. Where others in her race seemed to struggle, to have lost their spring, she loped around the track with the panache of an Olympian. She finished in 3:28, a new Senior Games record—and a time that few forty-year-olds could equal—and everyone cheered. It was spectacular, even beautiful to watch. I had to find out her secrets.

Her name is Jeanne Daprano, and she is a seventy-six-year-old former third-grade teacher from outside Atlanta. She is well known in senior track circles, holding multiple world records for her age group. In 2012, she became the only seventy-five-year-old woman ever to run a sub-seven-minute mile; according to the age-grading formulas used in Masters athletics, her 6:58 time translates to a young person running the mile in 4 minutes flat, which no woman has ever done. (The current female world record is 4:12.)

She was, in short, a ringer—yet her life story is hardly that of an elite athlete. She grew up on a farm in Iowa, and for most of her life her only real exercise consisted of keeping up with her elementary school kids in Long Beach, Southern California. She had taken up jogging at forty-five, during the running boom of the 1980s, but only got serious about training and competing when she turned sixty. "The women I started running with years ago were better athletes than I was—but they're not running anymore," she told me.

As a teacher, she would take her kids to running meets. Now she does nearly all her training runs on soft grass, to preserve her knees (which, obviously, were not really designed to last seventy years); and like many athletes she pays careful attention to her diet. As she got more into running, she gave up French fries—which she loved—for six weeks, just to see if she could. At the end of six weeks, she was cured of her French-fry addiction, and she now thrives on a diet of "living foods," by which she means raw foods, salads, and sashimi. She spends her spare time visiting nursing homes, trying to motivate the residents—some of whom are younger than she is—to stay active, and eat more healthy food.

"Aging is beautiful," she told me as we stood there in the infield. "Put in there that it's beautiful. God's design for aging is perfect."

It was hard to argue the point with her. But the question

occurred to me: Did Jeanne Daprano perhaps have more in common with Irving Kahn, the 109-year-old investment guru, than with her old running buddies? Were Jeanne and Ron and Don and Bernie and their fellow competitors also genetically protected from aging, somehow? Maybe God's plan for aging (or whoever's) is more perfect for some than for others.

Nathan LeBrasseur has the same question, and it hasn't really been answered in the existing research. "The problem with these studies is you don't know if the capacity to exercise well, to have a high VO_2 max, or to be motivated, is part of same the genetic signature as longevity," he says. "If you have the ability to train and protect fitness, are those the same things that protect against cardiovascular disease and stroke?"

We all know entire families of couch potatoes—but are they couch potatoes because of heredity, or because everyone around them is also a couch potato? Recent research suggests that the very willingness to exercise may be at least partly inherited. Studies of twins, for example, have found that close genetic relatives maintain similar activity levels throughout life. That doesn't really answer the heredity/environment question through.

Scientists at the University of Missouri recently tried to parse the issue with an interesting experiment: They separated a bunch of lab rats into two groups, those who ran enthusiastically and those who wouldn't go near a treadmill. They then selectively bred the runners with other runners, and the lazy rats with lazy rats. Within eight generations, they found distinct differences in the brains of the two lines of rats. The running rats had more of a particular type of neuron that is related to pleasure and addiction, which meant they were more likely to derive pleasure from exercise. The couch-potato rats had fewer of those neurons, which are located in a brain region called the nucleus accumbens. But then the researchers added a twist: When the couch-potato rats were induced to run (via electric shock), they too grew more of the "exercise neurons." So even the genetically programmed couch-potato rats learned to like it, at least a little.

This is an important finding, since studies indicate that up to 90 percent of Americans fail to meet the minimal guidelines for physical activity, defined by the federal government as thirty minutes of moderate exercise (think "brisk walking"), five times per week. As we engineer the activity out of life with each passing generation—en route, perhaps, to a La-Z-Boy-bound world like that depicted in the movie *Wall-E*—humans in the developed world tend to get lazier and lazier, less willing to be active (and rarely required to be), more like the couch-potato rats.

But at the same time, both the willing and unwilling rats were still better off doing something than doing nothing—and the same holds for people. Researchers are finally getting funding for clinical trials of exercise as an "intervention"—that is, evaluating exercise as if it were a drug—and one of the largest, the LIFE Study, reported results in June 2014. In the experiment, a group of about eight hundred sedentary older people in their seventies and eighties were induced to begin a mild exercise program—presumably not via electric shock, but who knows?

They were already in trouble, scoring very low on physical performance tests, but they could at least still walk a quarter mile, considered the threshold for something called major mobility disorder. Even so, these were the sorts of folks who were unlikely to go and join the gym; they were on the brink of no longer being able to live independently. After two years, the exercise group had far lower rates of disability than a matched group that had merely been *told* they should exercise. A little bit of walking kept many of them out of the nursing home, at least for a bit longer. If it were a drug, it would have been a slam dunk for FDA approval.

"It really is our most promising intervention for the ills of late life," LeBrasseur says. And, it's free.

Chapter 11

STARVING FOR IMMORTALITY

I stumbled out of Harrisburg. Cursed city! The ride I proceeded to get was with a skinny haggard man who believed in controlled starvation for the sake of health. When I told him I was starving to death as we rolled east he said, "Fine, fine, there's nothing better for you. I myself haven't eaten for three days. I'm going to live to be a hundred and fifty years old." He was a bag of bones. A floppy doll. A broken stick. A maniac.

—Jack Kerouac, *On the Road*

Answering the door of his stately home north of Boston, Don Dowden greets me with the news that I've missed lunch. Guiding me into the kitchen, he shows me the big metal salad bowl, now empty save some traces of dressing as well as a few remnants of spinach greens, red peppers, broccoli, mushrooms, and a stray chickpea or two. "It was really good," he says, and having just driven 150 miles without any breakfast, I have to agree. It even smells delicious.

But even in my hypoglycemic haze, I can appreciate the irony of the situation: I'm standing there starving, while Dowden—a retired patent attorney who practices what Kerouac called "controlled starvation for the sake of health," also known as caloric restriction—is licking his chops. Normally, he's the guy who should be hungry. Immediately, the tall, elegant Dowden reminded me of another patrician figure who adhered to the same discipline, only this guy did it five centuries ago, before it was cool.

The other guy's name was Alvise Cornaro, and he was a wealthy merchant and property owner who lived in Padua, Italy, in the sixteenth century. He was a largely self-made man, and evidently, he loved to party. His friends called him Luigi. By the time he reached his late thirties, though, his fast living had taken a toll, and he was tormented by ill health: "colic and gout, an almost continual slow fever, a stomach generally out of order, and a perpetual thirst," as he confessed.

Phil Bruno and any other diabetic would instantly recognize these as key symptoms of diabetes, a thoroughly modern disease of aging. Although Johann Sebastian Bach is sometimes cited as the first recorded diabetic, poor Luigi Cornaro beat him to it by a couple of centuries. He was not yet forty, but he was in such discomfort that, when he was honest with himself, he had to admit that he would almost welcome death. He couldn't let that happen. His gorgeous wife, Veronica, had just borne him a long-awaited daughter, and he had to live to see her grow up.

His doctors instantly pinpointed the cause of his distress in his "intemperate" lifestyle—in short, too many Renaissance feasts. (We've all been there.) The learned physicians, citing the already-ancient advice of the Roman physician Galen, told him to cut back a little. Naturally, he ignored their advice and kept feasting with his friends. (Been *there*, too.) "This, indeed, like all other patients, I kept a secret from my physicians," he admitted later.

Eventually, though, his illness forced him to surrender. His doctors had warned him one last time that if he did not cease and desist, he would be dead within months. This time, he resolved to change his ways. He started by cutting out foods that disagreed with him: "Rough and very cold wines, as likewise melons and other fruits, salad, fish, and pork, tarts, garden-stuff, pastry, and the like, [which] were very pleasing to my palate, [but] they disagreed with me notwithstanding." That did not leave much, except maybe ice—but he cut that out, too. Talk about a tough New Year's resolution.

Over the space of a few months, he settled on a healthy but meager-sounding daily ration of bread, a little meat (goat or mutton), or some poultry or fish, usually mixed into a brothy soup, with an egg yolk to make it more filling. He allowed himself exactly twelve ounces of this potage every day. "I always rose from the table with a disposition to eat and drink more," he remarked, somewhat unsurprisingly. "I take but just enough to keep body and soul together."

Spare as it sounds, his new diet literally gave him a second life. He began feeling better within a week, which gave him the strength to continue. Rather than dying at forty, as his doctors feared, Cornaro went on to become one of the wealthiest and most important men in Padua, and eventually one of the oldest. Tintoretto painted his portrait, which now hangs in the Palazzo Pitti in Florence. In the picture, whose year is unknown, he is bald and obviously aged, but his eyes dance with life. He was a happy man. He boasted to a friend that his way of living "has given me the vigour of thirty-five at the age of fifty-eight."

That was only the beginning. Even in his eighties, he was still bounding up and down the stairs of his estate, and working in his gardens. "I can mount a horse with ease," he bragged, "and many other things."

Not everyone was so thrilled. His worried family urged him to eat just a little bit more. He bumped up to fourteen ounces per day, but complained that the extra food made him "melancholy." So he went back to his twelve ounces, and they quit bugging him. He could return in peace to his main work, drafting a long-in-the-making treatise about his new lifestyle. He titled it *Discorsi della vita sobria*—Discourses on the Sober Life—and the first edition appeared in 1558, when Cornaro had reached the grand old age of eighty-one, double what his doctors had predicted for him. He revised it two years later, at eighty-three, and then again at age ninety. As he continued into his nineties, still in perfect health, he felt the need to rework and extend it yet a fourth time, at the ripe age of ninety-five—proof, perhaps, that the only thing that ends the editing process is death itself, to which he finally succumbed at age ninety-eight.

In all four versions, though, the basic message was pretty simple: *Don't eat so much.* A man, he wrote, should eat "no more than is absolutely necessary to support life, remembering that all excess causes disease and leads to death."

Cornaro's earnest little tome went on to become the world's first best-selling diet book. Strikingly modern in its confessional tone, it was translated and republished in nearly every language and in every century up to our own, becoming one of the most popular works on diet and longevity that has ever been written. It appeared in German, French, and English, including an edition compiled by none other than Ben Franklin, who no doubt recognized its commonsense appeal.

Cornaro's work enjoyed a renewed vogue around the turn of the twentieth century, when the growing temperance movement misinterpreted his use of *sobriety*—conveniently forgetting that he washed down his twelve ounces of soup with fourteen ounces of wine, or almost three glasses. Thomas Edison blurbed the book—now usually titled *The Art of Living Long* or *How to Live to 100*. Henry Ford handed it out to his rich friends, and new editions kept appearing all the way up into the 1980s. Through his book, Alvise Cornaro had earned the immortality he craved. But he would also make his mark on science.

One person who discovered the *Discorsi*, and scribbled in its pages, was a young professor of nutrition at Cornell named Clive McCay, who had come across a 1917 edition published as *The Art of Long Living*. Intrigued, McCay tried putting some lab rats on the Luigi Cornaro diet. He fed one group of baby rats a reduced-calorie diet (enriched with vitamins, to avoid malnutrition), while the rest were fed normally. Though the underfed animals remained smaller and scrawnier (or in his word, "retarded"), they lived nearly twice as long as their tubby cousins—up to four years, in some cases, which is a long time for a rat.

McCay's resulting paper, written with his graduate student Mary Crowell and published in the *Journal of Nutrition* in 1935, is now considered one of the great breakthroughs in our understanding of aging—but at the time, other scientists regarded their work as a bit odd, and of course irrelevant. Nutrition, in those days, was a bit of a backwater: "People hardly rate it a science at all," he complained. And if nutrition was a scientific backwater, then the study of aging—gerontology—was practically a desert island.

The public, on the other hand, was fascinated. In a radio talk, McCay claimed that his rats had lived the equivalent of 120 human years, which got people's attention. "The lifespan is probably much more flexible than we have imagined," he declared. He was written up in *Time*, and he gave countless radio interviews to a public that was intrigued by the idea that you could starve your way to a longer life. The Social Security program was brand new, and Americans wanted to be able to reap its bounty for as long as they possibly could. The Rockefeller Foundation soon came calling, with a $42,500 grant to further McCay's work on "diets that may promote longevity." Family patriarch John D. Rockefeller was ninety-six years old at the time, too late to benefit from McCay's work, and at any rate the various experimental diets he tried on the Rockefellers' dime—feeding the rats coffee, vitamins, "organ meats," and whole wheat bread—failed to inspire them to live as long as plain old hunger did.

The public's new enthusiasm for longevity research faded with the approach of World War II, which was shortening lifespans rather abruptly across the globe. Food and sugar rationing dampened any residual appeal of voluntary starvation, and McCay turned to other endeavors. Working with the military, he developed a dense, highly nutritious, and let's just call it *chewy* new kind of loaf that was dubbed "Cornell Bread"; it caught on, and could still be found in health-food stores throughout the country up into the 1980s. Little did people know that it was based on McCay's own special recipe for rat chow.

McCay never gave up his fascination with longevity, however. A gentle soul who had been orphaned at a young age, he surrounded himself with animals, particularly old dogs. His Green Barn Farm outside Ithaca, New York, was a veritable kennel, full of well-loved strays as well as retired beagles who had been used in research. Their scientific careers were not entirely behind them, as McCay kept fiddling with their diets. His research helped establish the basic nutritional requirements for modern dog food. And to his utter non-surprise, he found that dogs on a limited diet seemed to live healthier lives than those who were fully fed. (Advice I took to heart as I fed my own mutts—always at the lower end of the suggested feeding range.)

Someday, he would apply his research to humans, if he ever got the chance. "We have learned to keep most of our children from dying but we have not made much progress toward giving men and women a longer and healthier middle age," he lamented, toward the end of his life. "We do not wish to prolong the suffering that goes with feeble old age; we want to extend the prime of life when most of us live and enjoy living."

After Clive McCay's death, the gospel of starvation for the sake of health found an unlikely apostle in, of all places, Venice Beach—in, of all decades, the hedonistic 1970s. Part hipster, part artist, part Svengali, and part scientist, Roy Lee Walford had long been fascinated by aging, even as a teenager in San Diego in the early 1940s. "As a young man, he just wanted to live forever," one of his (many) girlfriends told the documentary filmmaker Christopher Rowland in 2007. "If you extended your life, you could have multiple careers, you could have multiple marriages—you could accomplish all kinds of things in this world."

Walford wanted to do it all. He was an actor, a writer, and an adventurer, ahead of his time in many ways. Also smart: He paid off his student loans, according to one story, by going to Reno with a mathematician friend and hanging around for three days analyzing how the roulette wheels were rigged. Then they bet accordingly, and cleaned up. He took periodic breaks from science, including a year spent traveling around India in a loincloth "as a naked seeker," as he put it, carrying a thermometer in order to study body temperature and aging in Indian yogis. Back home in Los Angeles, he palled around with a fast crowd that included Timothy Leary and members of the Living Theatre group. He stood out from the long-haired Venice hippies with his shaved head and blondish-gray Fu Manchu facial hair—again, ahead of his time. He got into punk rock before most of the rest of the world even knew what it was. He was in his fifties by then. "He was a bit of a wild man," recalls Rick Weindruch, his graduate student and protégé. "He lived life on the fringe."

Walford became fascinated by Clive McCay and his starved rats. McCay and others had believed that caloric restriction was extending the animals' lives by "retarding" their development from birth; slower growth equaled slower aging, they thought. Walford and Weindruch suspected that caloric restriction might actually be slowing the aging process itself, on a more fundamental level. They proved it by taking adult mice and gradually cutting back their food supply. It had been thought that older animals could not survive a reduction in food, but in fact they lived longer, with a much lower incidence of cancer. The restricted diet seemed to "shift them to a different metabolic and physiologic state, consistent with slower aging," Weindruch says.

They had no idea how this happened, but it was enough to convince Walford to get serious about his own diet: He cut down to a protein shake for breakfast, a salad for lunch, and a baked sweet potato and perhaps a bit of fish for dinner. He stuck to this menu, which Cornaro would have considered gluttonous, pretty much for the rest of his life. And he began preaching the wonders of caloric restriction to anyone who would listen. In health-conscious California, he found a receptive audience. He published a series of popular books, beginning with *The 120 Year Diet*, which sold well despite its title. "Who wants to be on a diet for 120 years?" Weindruch asks.

But he hardly denied himself. "Roy was very different from most other caloric-restriction practitioners who I've met," says Weindruch. He kept himself rigorously fit by lifting weights at the original Gold's Gym in Venice Beach. At dinner parties, he would conspicuously abstain from eating while everyone else pigged out; he joked to a friend that this made him feel "naughty." Supposedly, he only starved himself every other day; on alternate days, he was known to inhale stunning quantities of food. His friend Tuck Finch noted that their raucous, boozy dinners together always seemed to fall on Walford's "on" days, which still strikes Finch as "statistically unlikely."

Thanks to Walford, and McCay before him, "controlled starvation for the sake of health" had attracted a small but committed group of followers, including Don Dowden. Scientifically speaking, though, the jury was still out on the question of whether or not caloric restriction was actually good for people. For obvious reasons—mainly because the study would take decades—there was no good data on human subjects.

But then, in the early 1990s, Walford got the chance to be part of something that would radically change his career, and his life. In one of his periodic restless phases, he signed on as chief medical officer for Biosphere 2, the famous (or infamous) earthbound "space station" that was being built in the desert north of Tucson.

"I find it useful to punctuate time with dangerous and eccentric activities," he explained to the *Los Angeles Times.*

Funded by the venturesome oil heir Ed Bass, who considered himself an environmentalist, Biosphere 2 was a glass-enclosed, 3.15-acre terrarium that was designed to replicate the major ecosystems of earth (Biosphere 1). Walford and seven other "Terranauts" would spend two years inside the hermetically sealed chamber, living off the food they produced in their extensive organic gardens and indoor fish farm. They would receive nothing from outside, not even air or water, which would be recycled by the indoor ecosystem.

When the crew entered the Biosphere on September 26, 1991, Walford cut a striking figure in his *Star Trek*–style uniform, which perfectly matched his Spock-like ears and shiny dome. Things soon took an unexpected turn, however, when the explorers discovered that they could not produce enough food to feed themselves. Spotting an opportunity to turn lemons into diet lemonade, Walford decided that this was the perfect chance to study caloric restriction in people: Henceforth, the eight crew members would be placed on a sharply reduced ration of less than eighteen hundred calories per person per day, at first. As team physician, Walford would monitor its effects on them.

Normally, humans are hardwired to cheat on any kind of diet, which is another reason why it is so difficult to study caloric restriction. But now the Biosphere had presented Walford with eight captive lab rats, for two years. Their so-called healthy starvation diet was heavy on fruits (they grew bananas, papayas, and kumquats), and a long list of vegetables, nuts, and legumes, plus a handful of eggs, dairy from their goats, and a very small amount of tilapia and chicken. Only 10 percent of their calories came from fat, and they ate meat only on Sundays. All this was supposed to fuel them through eighty-hour weeks of serious manual labor, including tending the crops, maintaining heavy equipment, pruning back vines that climbed the glass-steel walls, and even donning scuba gear to clean the fish tanks.

Not surprisingly, the Terranauts lost weight like sumo wrestlers in a steam room, shedding pounds until their average BMI dropped below twenty for men and women alike (or in scientific terms, "really skinny"). One man lost 58 pounds, going from a portly 208 to a sleek 150. They lost weight so fast that Walford grew concerned that their fat cells were releasing toxins, like pesticides and pollutants, back into their bodies. They were indeed, he found, but the strict diet and heavy physical workload also caused more immediate problems, like that they were starving. According to crew member Jane Poynter, who wrote a memoir revealingly titled *The Human Experiment: Two Years and Twenty Minutes Inside Biosphere 2,* it became accepted practice to lick one's plate clean after every meal, so as not to miss a single precious calorie. The supply of bananas, the tastiest item on the menu, had to be kept under lock and key. Saddest of all, the Terranauts would occasionally peer through binoculars at tourists eating at the on-site hot-dog stand, like monks watching porn. "Roy was having a whale of a time, however," Poynter deadpanned, "because this was his life's work."

This was true. Walford had only ever been able to observe CR's effects on laboratory mice (also rats, fish, and monkeys). Now he was able to measure its effects on humans, including himself. He sampled the crew members' blood every eight weeks, and found that they had the best blood he had ever seen. Their cholesterol levels had dropped drastically, from an average of more than 200 down to well below 140. Their insulin and blood glucose levels also plummeted, as did their blood pressure, according to a paper Walford published while still "inside." Metabolically and cardiovascularly speaking, these were some of the healthiest people on the planet. Or so they seemed.

When the eight Terranauts emerged from the Biosphere, in September 1993, pomp and ceremony competed with sheer relief that the long, intensely scrutinized project was finally over. Though it had begun in an atmosphere of gee-whiz optimism—this is how we'll live on *Mars*!—the project had endured withering skepticism and a spate of negative press, including a *Village Voice* takedown that explored the project's roots in a strange organization called Synergia, which the paper characterized as a cult. Two years of confinement had divided the crew into bitter, warring factions; the tension and drama inside the Bubble actually helped inspire the reality TV series *Big Brother*. The meager diet had not helped morale, either. Opening the "seal" was meant to be a joyful day for all concerned. At least now they'd be able to visit the hot-dog stand.

For Walford, though, the end of the Biosphere marked the beginning of a dark new chapter of his life. He had been fit and vibrant when he entered the capsule, looking far younger than his sixty-seven years. Two years inside had ravaged his body. Perhaps it was the lack of food, perhaps something else, but in photos taken in the Biosphere, Walford is thin to the point of emaciation, his eyes haggard and sunken. He'd lost 25 pounds from an already-lean 145, and he looks much older than the post-Biosphere version of himself, on the right.

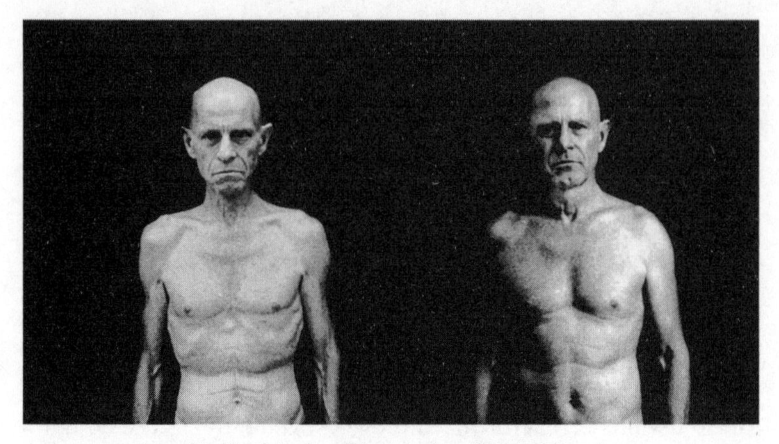

Roy Walford during (L) and after (R) the Biosphere.
Credit: *Journal of Gerontology*

But the real damage was invisible. In the six months after leaving the Biosphere, Walford fell into a deep depression, drinking his way through a bottle of vodka every four days. He had injured his back while working in the compound, and he could barely walk, at times. Something seemed to have changed in his brain, as well: Just three years after leaving the Biosphere, he began experiencing episodes of "freezing," where he would simply stop walking, and fall down. Soon he required a walker.

Those close to him suspected that he had acquired some form of Parkinson's, and that it may have been caused not by caloric restriction, but by oxygen restriction. The Biosphere's designers had not foreseen that the vast concrete surfaces in the complex would absorb literally tons of precious oxygen, in effect asphyxiating the crew. Walford himself grew alarmed, about six months in, when he found he could no longer perform simple calculations.

Even after the atmosphere was "rebalanced," by injecting oxygen and installing more CO_2 scrubbers (which journalists later gleefully discovered), oxygen levels remained low. The Biospherians were living at the atmospheric equivalent of seven thousand feet above sea level, which can be taxing. Not only that, but CO_2 and carbon monoxide concentrations crept dangerously high. Carbon monoxide exposure, in particular, has been linked to Parkinson's and other neurological disorders.

Despite his illness, though, Walford's mind remained sharp, and he stuck to his diet, insisting that it had slowed its progression, rather than hastened it. As late as 2001, he touted the benefits of caloric restriction to Alan Alda, who then hosted a TV show for *Scientific American*, insisting that caloric restriction would "let me live longer than I would otherwise." He said he hoped to live to be 110 years old, just like Suzanne Somers.

But physically, he was a wreck. A video taken of Walford that same year is truly shocking: Barely a decade after entering Biosphere, as a vigorous real-life version of Captain Jean-Luc Picard, he had been reduced to a stooped, jittering old man, hunched over and barely able to walk on his own. He had already been diagnosed with Lou Gehrig's disease, from which he eventually died in 2004. Although immortality had eluded him, Walford had packed more living into his seventy-nine years on earth than most of us could fit into three lifetimes.

By the time Roy Walford died, he and other scientists had spent decades studying caloric restriction—make that centuries, if we count Luigi Cornaro—without ever understanding one important thing: how it actually worked.

The most surprising thing about caloric restriction was *that* it worked, period. It goes against all common sense; you'd think that starving animals would grow weak with hunger and perhaps even die. But in fact the opposite is true. Mice placed on a restricted diet become far *more* active, running literally miles farther, when their cages are equipped with treadmills, than their normally fed brothers and sisters, according to Rick Weindruch. It's not only mice and rats, either; a wide range of creatures, from dogs all the way down to lowly yeast, have all been shown to fare better on restricted diets. And for the longest time, nobody had the faintest idea why.

Weindruch and Walford suspected that lack of food somehow shifted the animal—whether it be a single-celled yeast, or a mouse, or a human—into a different metabolic state that was, somehow, healthier. "It isn't that this car is blue and this car is red," says Rozalyn Anderson, a colleague of Weindruch's at the University of Wisconsin. "It's that this car has a different engine."

In the early 1990s, an MIT scientist named Leonard Guarente discovered a specific gene in yeast that seemed to respond to lack of nutrients. The one-celled booze makers seemed to be able to sense the amount of food that was available to them, and reprogram their metabolism accordingly, in ways that helped them live longer. The actual gene itself was called SIR2, and it seemed to be responsible for optimizing cell function in response to a lack of food.

The story grew even more intriguing when SIR2-like genes were found in other organisms, from worms to fruit flies to mice to monkeys. These genes, dubbed sirtuins, appeared to be what scientists call "conserved"—that is, they had evolved in numerous different species of animals, and even in some plants. This meant they were somehow important to life itself, and it's easy to see why: They enabled animals to survive the long periods of hunger that are part and parcel of life in the natural world. A hunter-gatherer who could endure a lean winter, growing stronger and healthier while eating less food, would enjoy an evolutionary advantage over one who required a steady diet of Big Macs. To the contrary, it's when we're well fed and bloated that our genes seem to want to kill us off.

The discovery of sirtuins electrified the aging field. Their existence meant that, on some level, we had longevity pathways hardwired into our cells. Then, in 2003, a student of Guarente's named David Sinclair discovered that sirtuin genes could be activated by a compound called resveratrol—which just so happens to be found in red wine (it's produced in the skins of grapes, where its job is to ward off fungal infections as the fruit ripens). In a study published in *Nature* in 2006, Sinclair and his team showed that mice on a high-fat diet lived just as long as normal mice—when they were given resveratrol. Not only that, but they were fitter, faster, and lots better-looking than their chubby colleagues.

The media went bananas over the story. It made the front page of the *New York Times*, and now it was Sinclair's turn to clink glasses with Morley Safer of *60 Minutes*. He dazzled Barbara Walters with his youthful good looks, and within days, it seemed, the Internet was flooded with ads for resveratrol supplements, some of which implied that Sinclair himself endorsed their products (which he did not). One of the few existing resveratrol supplements, a brand called Longevinex, saw demand soar 2,400-fold in the space of two weeks. If ever there was a drug tailor-made for overweight, fast-food-gorging Americans, resveratrol seemed to be it.

And it dovetailed nicely with the so-called French Paradox,

where the French eat all sorts of fatty foods and yet do not succumb to heart disease (or obesity) at nearly the rates that Americans do. Red wine was long thought to be the reason, and lo, resveratrol is in red wine; clearly, something in red wine is good for you. Maybe several somethings. (Scotch and beer drinkers, and even white wine drinkers, also enjoy a health boost over teetotalers—but do not benefit quite as much as those who consume red wine.)

Meanwhile, research on resveratrol became its own mini-industry, as hundreds of papers were published on the miracle pill that seemed like it might slow down the aging process. It also apparently boosted endurance, causing some members of the Sinclair lab to take it in search of an athletic performance boost. Critics were few and far between, but vocal. Indeed, some of his leading antagonists included fellow alumni of Leonard Guarente's lab, which must have made reunions awkward. (For more on resveratrol as a supplement, and why it might not actually be the miracle pill it was touted as, see the Appendix, "Things That Might Work.")

It didn't matter; Sinclair had already moved beyond resveratrol to much bigger things. About a year and a half after the *Nature* paper, GlaxoSmithKline plunked down $720 million to buy Sirtris, a startup pharmaceutical company that Sinclair had cofounded. The company was supposed to go on to develop sirtuin-activating drugs that would be better targeted and more specific than resveratrol, which Sinclair dismisses as "a dirty drug." Sirtris's new drugs would also be patent-protected, which meant that a drug company could make money from them.

But then . . . crickets. GSK/Sirtris brought a handful of new drugs into clinical trials, but they fared poorly. One had to be withdrawn because of side effects, and in March 2013, GSK shuttered its Sirtris office in Cambridge, Massachusetts. Although the company said it had not abandoned the cause completely, only a handful of staffers seemed to be still working on the project.

"It's a business decision, and not because of the science," Sinclair told me, a week after GSK had pulled the plug. He hadn't given up, but he sounded resigned. "I can remember telling my friends in college, we were playing cards, and I said, 'Do you realize we're probably the last generation that's going to live a normal lifespan? Someone's going to make a breakthrough, and the next generation's going to live a long, long time. And that sucks, because we were born one generation too early.' "

Another reason it "sucks" is because, in the absence of a caloric-restriction pill (which is how resveratrol was billed), then we apparently still have to starve if we want to live longer. But how much hunger does it really take?

To help answer that question, I had sought out Don Dowden. I'd found him on my search for a modern-day Luigi Cornaro. Tall and patrician, he had been a successful patent attorney in Manhattan for decades before retiring to this place north of Boston, which seems to be on a sort of family estate. "He's interesting," a friend had said. "He's older, but he's very lively."

I saw instantly what he meant. Dowden is in his early eighties now, and while his skin shows his age, his eyes dance with alertness. We sat down on the shabby-genteel furniture in his study, and he told me his story: When he was still in his late twenties, just married and starting out in New York, he happened to read a report by the American Society of Actuaries "that said it was better to be thin than fat," he recalls.

This was in 1960, when such insights were still novel, and they hit home: He'd gained thirty pounds immediately after getting married, thanks to his wife's cooking. He decided to try to take the weight off, and his plan was simple: He'd eat less. He also started jogging, which was regarded as a bit eccentric back then—and actually, so was the notion of eating less, particularly in the steak-and-martini milieu of Midtown Manhattan in the early 1960s.

Then, in the early 1980s, he read Roy Walford's bestseller *The 120-Year Diet*, and he resolved to be more systematic about it. He joined the Calorie Restriction Society, a group of like-minded people (including Walford) who enjoyed really tiny meals. He soon saw why its membership numbers have never been particularly robust: He threw a party for fellow CR practitioners, and one guy brought a scale to weigh the food. "I weigh myself, but I don't weigh the food," he says.

He certainly seems youthful for his age. At eighty-two, he takes no medications and is proud of the fact. He carries 155 pounds on a six-foot-one-inch frame, less than his college weight, he humblebrags, but he does not look emaciated. He kept running into his seventies, when his knees finally gave up on him, and he still goes out for a daily walk in the woods for an hour or ninety minutes. All that's missing is his wife, who died in 2000 of a neurological disorder at age sixty-five; she had not joined his caloric-restriction diet, he says.

Was all this self-deprivation worth it? Dowden thinks so. In the mid-2000s, he volunteered for a study led by Luigi Fontana, a scientist at Washington University in St. Louis who is the leading researcher into caloric restriction in people (and also a practitioner of CR himself).

Dowden was one of thirty-two members of the CR Society who participated in the study, and just like the Biosphere inmates, the

CR practitioners had lower blood pressure, better cholesterol, and much healthier arteries than the controls. This was expected, but they also turned out to be much healthier than a group composed of regular marathon runners. That was surprising. By these measures, Fontana concluded, Don Dowden and his fellow undereaters tested decades younger than their chronological age.

And the world was all like, *Big deal.* Just another strange American dietary subculture, along with the vegans and fruitarians and raw-foodists and juicers and all the rest. The rest of us regarded them as floppy dolls, broken sticks, maniacs.

Until July 9, 2009, that is. On that day, two very special monkeys made the front page of the *New York Times.* They were pictured side by side: Canto, age twenty-seven, and Owen, twenty-nine. In monkey terms, this made them the equivalent of senior citizens, but the striking thing was that Owen looked like he could have been Canto's beer-drinking, dissipated dad. His hair was patchy, his face

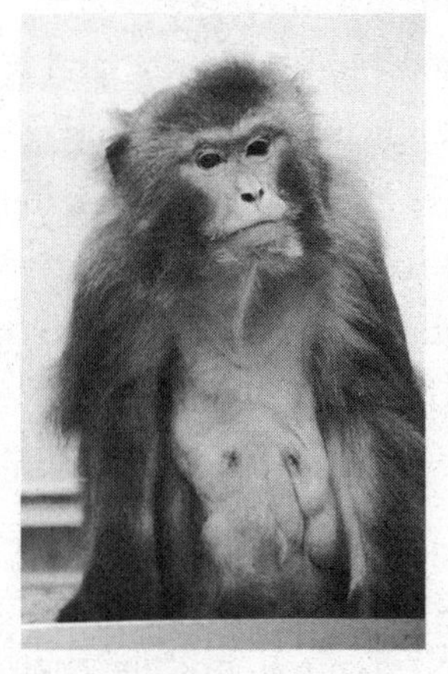

Canto, left, was on a reduced-calorie diet his whole life;
Owen, right, was not. Obviously.
Courtesy University of Wisconsin Board of Regents.

sagged, and his body was draped in rolls of fat. Canto, on the other hand, sported a thick (if graying) mane, a slender physique, and an alert, lively mien—like a simian Don Dowden.

The two monkeys were part of a long-running study of dietary restriction and aging, conducted at the University of Wisconsin and spearheaded by Rick Weindruch and Roz Anderson. Since late adolescence—sometime in the late 1980s—researchers had been feeding Canto and a few dozen other unfortunate monkeys about 25 to 30 percent less food than Owen and his well-fed crew. The monkeys were primate stand-ins for humans, close enough to draw conclusions as to whether caloric restriction would really delay aging in people. (An early director of the National Institute on Aging had proposed running the experiment on prison inmates, but his idea was vetoed.)

The study lasted decades—lifespan studies are nothing if not tedious—but by late 2008, the scientists could report meaningful results. The differences were as striking as the side-by-side photos: The hungry monkeys were far healthier, in terms of basic measures like blood pressure, and had far less incidence of age-related disease, such as diabetes and cancer. As a result they seemed to be living as much as 30 percent longer than their overfed friends.

In a word: Wow. (And in two: No, thanks.) Regardless, caloric restriction had now burst into the popular consciousness—at last, a cure for aging! The only bad news is that you can't eat.

The most striking finding in the Wisconsin monkeys, though, was revealed in brain scans.

The caloric-restriction monkeys had retained far more gray matter—a novel finding that had not been seen in mice on caloric restriction (or, obviously, in yeast). Like humans, monkeys undergo a long, steady process of brain atrophy as they get older.

The underfed monkeys seemed to be protected from this brain aging, however. In particular, they had preserved brain regions responsible for motor control and "executive function," the part of the brain where we make our important daily decisions (like: salad or cheeseburger?).

Among scientists who studied CR, the monkey study was considered ultimate proof that eating less would extend lifespan, for everyone, all the time. It had even worked in Labrador retrievers, in a study sponsored by Purina. Of *course* it would work in people. But then along came the Whole Foods monkeys, and they really screwed things up.

Two years after Canto and Owen made the front page, a scientist named Rafael de Cabo sat in his office at the National Institute on Aging, in a shiny glass building overlooking Baltimore Harbor, in a state of serious anxiety. He had been combing through data from a second major study of caloric restriction in monkeys, one funded by the NIA, and the conclusion was as shocking as it was inescapable: This time, the "dieting" monkeys had *not* lived longer. CR was supposed to be a slam dunk, but instead this study had failed, taking millions of federal dollars down the drain and creating an incipient PR nightmare.

Like the Wisconsin study, the NIH study dated back to the 1980s. "There's no quick way to do an aging study," says Roz Anderson of Wisconsin. When the data began rolling in, de Cabo saw not just one problem, but two: Not only were his calorically restricted monkeys not living longer, like they were supposed to, but the NIH "fat" monkeys were *also* living a long time—just as long as the Wisconsin calorically restricted monkeys. What was *that* about?

De Cabo and his coauthors dutifully published their findings in *Nature* in August 2012, and the headlines were predictable: "Severe Diet Doesn't Prolong Life, At Least in Monkeys," declared the *Times*. Not only did it appear to call into question the whole theory behind CR, it was close to an institutional disaster: The government had just spent roughly $40 million to prove that caloric restriction helped monkeys live longer in Wisconsin, but not in Maryland. But the headlines also masked a much more nuanced—and hopeful—story.

Even though the "hungry" NIH monkeys didn't live longer on average than the ones who ate whatever, they actually stayed healthier for longer. They had lower incidences of cardiovascular disease, as well as less diabetes and cancer—and when those ailments did appear, they did so later in life. Also, four of the NIH monkeys lived past the age of forty, making them the longest-lived rhesus monkeys ever. "I think it's one of our very interesting findings," says de Cabo. "We can have a dramatic effect on healthspan, without improving survival."

A native of Spain, de Cabo is an anomaly in the world of dietary restriction research, for the simple reason that he loves food. In a field dominated by skinny people like Luigi Fontana who weigh their salads, he has the robust physique of a chef; his dream is to one day open a restaurant. He loves a good Rioja, and makes the meanest paella this side of Seville. "I love to cook," he says. "Would I like to practice caloric restriction? I don't think so."

And in fact, maybe he shouldn't, anyway: A wealth of good epidemiological data points to the fact that it's better to be a little bit overweight (i.e., BMI of 25) than to be seriously underweight (BMI below 21), like Don Dowden; the most ardent practitioners of caloric restriction can drop down to BMI of 19, which is considered dangerous. The reason, according to Nir Barzilai and others, is that very skinny people may not have the fat reserves they need to survive an infection, particularly as they get older.

Perhaps because he is a foodie, de Cabo zeroed in on the main difference between the two studies, which was the animals' diets. The Wisconsin monkeys had eaten a "purified" monkey chow comprising processed, refined ingredients, while the NIH monkeys had eaten a different formulation made from more natural ingredients and whole foods, including fish meal and grains. The Wisconsin diet allowed researchers to control the nutritional content more precisely. But because the NIH monkeys were eating more natural ingredients, they were taking in more polyphenols and other random compounds that we now know may have health-promoting effects. "There are micronutrients, there are flavonoids, that change depending on the time of year," de Cabo says.

There were other differences. The Wisconsin monkeys got their protein from whey (that is, dairy), while the NIH got most of theirs from soy and fish. Because of different feeding regimens, the NIH monkeys ended up eating 5 to 10 percent less than the Wisconsin fatties, which meant that they, too, were on a very slight form of caloric restriction. And while the NIH monkeys got just 5 percent of their calories from sugar, the Wisconsin monkey-chow was packed with more than 30 percent sugar, by calories. "That's like homemade ice cream," says Steven Austad. "My guess is [the Wisconsin monkeys] didn't like the food, and they added all that sugar to get them to eat it." (Sort of the way food companies add spoonfuls of sugar to, say, "plain" yogurt.)

So in effect, the NIH monkeys were dining at Whole Foods, while the Wisconsin monkeys ate at the ballpark—brats, beer, and funnel cakes—every day for more than thirty years. "The NIH monkeys were really on a fish-based Mediterranean diet," says Luigi Fontana. Put that way, it's not surprising they had different outcomes. And no surprise, either, that nearly half the Wisconsin monkeys had diabetes or pre-diabetes, just like 25 percent of the U.S. population; by and large, those were the monkeys who had died young. But as Rozalyn Anderson of Wisconsin points out, "Our diet was a lot closer to what people actually eat."

For all its flaws, then, the Wisconsin monkey study remains one of the best studies of the effects of a junk-food diet that has ever been done—and, clearly, the less junk food monkeys eat, the better.

People, too. Nobody has attempted a clinical trial of a fast food diet, but Luigi Ferrucci, the NIA scientist who runs The Blast (and a colleague of de Cabo's), recently tried an interesting and similar little experiment. In a small unpublished study, Ferrucci fed a hearty fast-food lunch to a couple of dozen volunteers, and then monitored their blood chemistry for the rest of the day. The fast-food eaters had extremely high levels of IL-6, the primary marker for systemic inflammation, for hours longer than a control group who had eaten a healthy meal of greens and salmon. It was as if the fast-food eaters had been physically injured by their food.

So, clearly, *what* you eat is as important, or more, than how much. Just as Hippocrates—who said, "Let thy food be thy medicine"—had worked out more than two thousand years ago.

Which is basically what Don Dowden does: His Cornaro-like diet is limited but far less austere than that of many food-weighing, calorie-obsessed CR devotees. Dinner is a piece of fish and vegetables or a salad (he eats a lot of salads), and he does not deny himself the occasional glass of wine. But he's flexible: If you invite him to dinner and serve him a steak burrito, he'll politely and WASPily eat at least half of it.

For the most part, though, he eats like an NIH monkey: not too much, and nothing processed, only good, real food. He looks forward to his salad lunches and his salmon dinners, as much as any gourmand. In fact, when I visited, he was getting ready to renovate his kitchen. And as I found out, he doesn't leave a lot left in the bowl. Is his diet responsible for his good health? Who knows; the answer may just as well lie in his genes, particularly those from his great-grandfather, who lived to be ninety-seven years old. But the regime works for him, and it's not too difficult, so he sticks to it, just like Luigi Cornaro.

"I bet there are millions of people in America who are basically doing what I'm doing," he says. "They just don't say they're 'calorically restricting'; they're 'on a diet.' "

And with that, he sends me on my way—to the nearest hot, juicy, New England–style grinder sub.

Chapter 12

WHAT DOESN'T KILL YOU

I'm not afraid of death; I just don't want to be there when it happens.

—Woody Allen

At a little past eleven on a Saturday morning, two middle-aged men strode onto a sandy California beach, took off their shirts, and ran into the breaking waves. This wouldn't have been all that remarkable, except that this was April in Half Moon Bay, just south of San Francisco, and the water temperature hovered somewhere below fifty degrees.

The air was even colder, with the chilling effect of a stiff onshore breeze and the fact that the sun was hidden behind the usual layer of damp coastal fog. At least the beach wasn't crowded. One guy played Frisbee with himself, tossing the disk toward the open sea and letting the frigid wind bring it back to him. The lone surfer was snuggled in a thick, toasty neoprene wet suit as he tried to coax a ride out of the two-foot waves. The sensible people were bundled up in hoodies or huddled in windproof tents.

Not these two idiots. They strutted onto the sand in Hawaiian surf trunks, not minding the stares their pale, cottage-cheesy flesh attracted. The older one, with thinning silver hair, bounded into the surf, chest-breaching two waves before diving into the face of a third with a whoop. The other guy, whose thinning hair was blondish, followed close behind—but then suddenly froze in place, in navel-deep water, where he let out a girlish shriek before his lungs clamped shut and he felt like he couldn't breathe, and was possibly about to have a heart attack right on the spot.

He then turned around and started running back *out* of the water, like a giant sissy.

That second guy, you might have guessed, was me. The first, fearless one was Todd Becker, a mild-seeming fifty-seven-year-old biochemical engineer who may just be the world's toughest nerd. Becker loves cold-water swimming and other painful-but-character-building activities, because he follows a school of ancient philosophers known as the Stoics, who believed that suffering breeds strength. All but ignored in the modern Western philosophical canon, the Stoics' theories have echoes in Eastern religions, notably Buddhism, which teaches calm in the face of struggle.

Becker uses their teachings as a guidebook for life. He is the leading, and perhaps only, practitioner of what he calls a "hormetic" lifestyle, in which he actively seeks out stressful experiences (like diving into bone-freezing water). It comes from *hormesis*, an ancient Greek word now used by scientists to describe a certain kind of stress response that has been observed in nature. The basic idea of hormesis is that certain kinds of stress or challenges—even some poisons—can actually elicit beneficial effects in the right doses. "We know that chronic stress in humans is clearly bad for you," says Gordon Lithgow, who has studied hormesis in *C. elegans* worms. "But short periods could be beneficial."

The phenomenon is observed in all kinds of organisms, from mammals on down to bacteria. The life-extending effects of caloric restriction are likely due to a hormesis-type response. In our own lives, we see it most often when we exercise, particularly lifting weights: The work stresses and even damages our muscle fibers, but thanks to the miracle of hormesis, we repair and rebuild them with new, stronger fibers. Most vaccinations work by the same principle: A small dose of a pathogen stimulates a response that renders us immune to the disease.

"Stress is strengthening, even essential to life," Todd had opined on the ride over from Palo Alto, where he lives. "Without it, we'd just dissipate away into nothing."

Which explains some of his rather odd daily practices. For example, every day he spends half an hour or more doing eye exercises, which he claims have strengthened his vision so much that he was able to toss out the Coke-bottle spectacles he used to wear. Each morning he wakes up to an ice-cold shower, the cold tap cranked all the way up. He stays in for five minutes, minimum, and claims that this bracing ritual not only jolts him awake (thus reducing his coffee budget) but also burns fat, improves pain tolerance, and boosts immunity. His essay about cold-water showering remains the most popular post on his well-read blog, gettingstronger.org. (Motto: "Train yourself to thrive on stress.")

The World's Toughest Nerd first dabbled with cold showering more than ten years ago, and it quickly became a cornerstone of his Stoic lifestyle. He swore that it had even helped him with his depression, for physiological as well as psychological reasons. "It makes it easier to cope with other stresses and situations in life," he told me. Incredibly, he had even convinced his son and daughter, ages nineteen and twenty-two, to try it. To his surprise, they actually liked it, though he'd had less success with his wife, who'd replied, "Absolutely no way!" End of discussion. He did manage to persuade a few friends to join the club, one of whom later remarked, "Cold-water showering is the scariest non-dangerous thing I've ever done."

Well put. Todd's other painful-but-non-dangerous activities include short-term fasting—skipping one, and sometimes even two meals—and tough, intense workouts, either running barefoot on trails or climbing in a nearby rock gym, often on an empty stomach. As a matter of fact, he hadn't eaten for twenty-two hours, or since lunch the previous day, despite going for a tough postwork run with work colleagues in the Palo Alto hills, training for an upcoming 125-mile relay race. "One thing I *really* love," he confessed to me, "is fasted workouts."

Awful as that sounds, we'll see in the next chapter why working out while you're hungry might actually be a good thing.

This frigid open-water swim was going to be a rare treat for him, a special luxury that he had generously offered to share with me. His hero is a mad Dutchman named Wim Hof, who holds a Guinness world record for swimming fifty-seven meters under Arctic ice. Like Hof, he had trained himself to withstand cold water for long periods. "The thing about cold-water swimming, or showering, is that the first time you do it is the worst," he'd said, as we rolled into the parking lot. "And then, the first minute is even *worse*. But after that, it gets better. It may sound strange, but it's true."

I gave myself thirty seconds, tops. I'm a guy who flinches at the thought of getting into the eighty-degree pool at the Y. A nice, hot shower is often the emotional high point of my day. Also, I couldn't help thinking about the fact that a cold-water swimmer had recently perished in nearby San Francisco Bay, at the start of the annual Escape from Alcatraz Triathlon. He'd died almost instantly, of what race organizers called a "massive cardiac event." He was forty-six years old, just like me. He probably thought he was fit, like me. And he had been wearing a wet suit. Unlike me.

We walked down to the beach and laid down our towels and gear. I still had my sweatshirt on, my last cozy security blanket. It felt at least ten degrees colder here than in the parking lot. Todd eyed the sweatshirt. "Now or never," he said.

"I'm good right here," I replied, digging my toes into the warmish sand and trying to forget something he'd said on the ride over: "I think the water is about as cold right now as it ever gets."

Instead, I tried to refocus on the delicious lobster rolls we would be eating later, at Sam's Chowder House. With some fries and a nice, warm beer...

"You only live once," Todd said, intruding on my reverie.

Or something like that. I don't really remember everything that happened next, except that I zipped off the hoodie, said "*Screw it!*" and took off running for the water's edge.

A human being exposed to cold water can die in one of two ways: quickly, or slowly. The former generally involves the cold-shock response, where blood pressure and heart rate suddenly skyrocket, resulting in a heart attack (or, as the press release will put it, a "massive cardiac event"). This is especially popular among middle-aged triathletes, and it is infinitely preferable to the slow way, where the cold water simply sucks heat out of your body—twenty-six times faster than when skin is exposed to air, by the way—until you eventually lose consciousness and drown. According to the Coast Guard, that generally happens when one's core temperature drops to below eighty-seven degrees, or about forty degrees warmer than the water in Half Moon Bay.

So I needed to be convinced that this cold-water plunge was going to be worth it, and not merely an "invigorating" exercise that I would hate, if it didn't kill me. Were there other benefits to short-term cold exposure—or was Todd just some kind of masochist? You know, the kind who deliberately takes a cold shower every day, and enjoys it?

I didn't have to dive very deep into the database to find some hints that a shot of cold might actually be good for you. Some of the longest-lived creatures on earth thrive in very cold water, for one thing. Lobsters can live for decades, if they are clever enough to stay out of lobster pots. Blue whales have been known to live for 200 years or so, and then of course there was Ming, the 507-year-old Icelandic ocean clam, may she rest in peace. Cold water may also explain why certain species of rockfish (a family that includes snapper and striped bass) essentially appear to be ageless, showing no signs of age-related deterioration at all. But whether cold water is responsible for their longevity, who knows?

Nematode worms, however, are pretty easy to keep in the lab, and studies on them suggest that cold water might help increase their longevity. For a long time, it was thought that this had to do with the simple fact that colder temperatures slowed down the chemical reactions that eventually lead to aging. New research, however, suggests that something much more profound might be going on. In a recent paper published in the journal *Cell*, scientists from the University of Michigan found that cold temperatures activated a specific longevity-promoting pathway in the worms.

The good news is that humans actually have the same genetic pathway. That may help explain why Katharine Hepburn, who swam every day in Long Island Sound, winter or summer, lived to the age of ninety-six. In a Finnish study of cold-water swimmers, subjects were analyzed at the beginning of the "winter swimming season" (yes, there is such a thing), and then again at the end, five months later. After months of regular cold swimming, the study subjects had much higher levels of natural antioxidant enzymes in their blood. They also had higher red-blood-cell counts, and more hemoglobin—in short, there was more blood in their blood. Their bodies truly had adapted to the stress of cold-water immersion, in good ways.

Even more interesting was research showing that exposure to cold water could activate brown fat, which burns energy and helps generate heat (for example, when we do stupid stuff like jump into frigid water). Unfortunately, brown fat is also relatively rare in adults, making up only a tiny fraction of our total fat stores; in most of us, it's limited to a few small pockets on our upper back, between our shoulder blades. It kicks into action in cold conditions in order to help keep us warm, but it also exerts beneficial effects on metabolism, mainly by burning white fat more quickly and efficiently than exercise or dieting. The more you are exposed to cold, studies have shown, the more brown fat you create. This is a good thing.

Finally, on a much more basic level, heat is bad for us; the fact that we are warm-blooded creatures actually accelerates our aging. It's part of the price we pay for not being reptiles. The reason has to do with our proteins—not the kind we eat in the form of meat or tofu, but the proteins that form the most basic currency of our cells. (Basically, everything that goes on in our cells depends on proteins.) The thing is, excessive heat causes our proteins to lose their three-dimensional structure, known as their "folding," which is crucial to their proper functioning. When they "unfold," then they can no longer do their jobs. We can observe this while cooking breakfast: "Crack the egg, it goes into frying pan, and the proteins unfold," explains Gordon Lithgow.

These "misfolded" proteins then become useless to our bodies, like empty wine bottles sitting around in our house. They have to be thrown out, or recycled, by a complex array of cellular garbage-disposal and protein-recycling machinery called the proteasome. But even at lower temperatures—say, around 98.6 degrees Fahrenheit—the same thing is happening: Our proteins are literally cooking.

This is thanks to a phenomenon called the Maillard reaction (named for the French guy who discovered it a hundred years ago), which we have all observed: It's why bread crusts are brown, why steaks get charred on the grill, and why cooked rice smells good. The Maillard reaction happens when amino acids combine with sugars under heat, and it makes food yummy. Unfortunately, the same reaction is taking place inside our bodies, a kind of slow-motion Maillard that produces nasty things called advanced glycation end-products, or AGEs—which really do help promote some diseases of aging, notably macular degeneration but also atherosclerosis, diabetes, and Alzheimer's. AGEs in our blood vessels, we know, are a leading cause of stiffer arteries and higher blood pressure, for example.

Bottom line, then: My proteins were loving the cold water. So was my brown fat. And while other parts of me weren't enjoying it quite as much, my brain knew enough to trust that Todd had done his homework and was maybe on to something. So I vowed to do my best to stay in that F*7@ki*g freezing Half Moon Bay as long as I possibly could.

* * *

It burned, at first. At least, that's what it felt like: wading into hot lava. That sensation didn't last more than a couple of seconds before the chilly reality hit. It was breathtaking, literally; I could not breathe. I panicked and stopped at about waist-deep. "*Can't breathe!*" I croaked, turning to bolt for shore.

Todd dove into a wave, ignoring me. I had no choice but to follow or get knocked down. When I surfaced, a strange thing happened: It got better. Sort of. The impulse to flee had faded, anyway. Another wave rose up, and we dove into that. I whooped for joy now; this was fun. Frisbee Guy had stopped his masturbatory throwing and was staring at us.

"It's important to move around a lot," Todd said—not to stay warm, mind you, but to make it even colder. "Otherwise you generate a little warm layer next to your skin," he added, like that was a bad thing. We jogged in place and body-checked the incoming waves, laughing and shouting like kids. My skin tingled all over; my legs were numb from the knees down. My heart pumped furiously, trying to keep my core warm, while robbing blood from my extremities. Like my testicles, which ached fiercely, as if in the grasp of a vengeful ex-girlfriend. "My balls hurt!" I blurted.

"Yeah," said the World's Toughest Nerd, with a knowing nod. "That happens."

The rest of me felt pretty good. A strange calm settled in, and I relaxed into the water, as if it were the same balmy Pacific that caresses Hawaii (which it is, technically). As I plunged into yet another hypothermic wave, I wondered, *Is this what happens when you die?* I almost felt too good. I looked at my watch, which said we'd been in the water for four minutes and change. It was cold, but I didn't mind so much. A short while later, a decent-size wave came along, and I bodysurfed it in to shore. I'd stayed in for almost six minutes, and, weirdly, I had enjoyed it immensely. It was the most exciting sort-of-dangerous thing I'd done all month.

"Nice job!" Todd said when he got out of the water after six more refreshing minutes. We toweled off, feeling like the biggest badasses on the beach, and got in the truck to go for lunch. A lobster roll had never tasted so good, but it was the thrill of the cold water that made it that way.

"It's like this gold mine, that people are not taking advantage of," he sighed over lunch. "It's so obvious to me, I don't understand it."

Back home, I tried the cold-showering thing a few times, telling myself that our city tap water wasn't nearly as frigid as Half Moon Bay. It would be easy! Also, the case for the benefits of cold water seemed pretty strong. And even if I did not end up living forever, it definitely started the day with a zing. And, often, a yelp.

It felt great after a hard bike ride on a hot day, or when I needed a bigger jolt of energy than yet another cup of coffee could provide. Once in a while. But my enthusiasm for it as a daily practice pretty quickly shrank, you might say. Standing under an ice-cold shower just wasn't as much fun as splashing around in the ocean. So while I will gladly jump into almost any cold body of water now, without fear or hesitation, in my daily life I went back to hot showering.

It was thus a relief to learn that just as cold can be good for you, small doses of heat are also beneficial in certain ways. Heat activates something in our cells called heat-shock proteins, which sound bad but aren't. Their job is basically to help repair cellular proteins, which tend to come unglued or unfolded in hot or stressful conditions—they help uncook the egg, you might say. Studies in smaller animals like worms have shown that this response can be "trained"—worms that have experienced stress in their lives develop a more powerful heat-shock response.

Heat-shock proteins are especially important in exercise. To compress a very long and complicated story: Certain heat-shock proteins, such as the ones produced in high-intensity exercise, actually help our cells clean house, which lets them function better for longer. Every time he cranked up the intensity in his Spinning classes, for example, Phil Bruno was creating heat-shock proteins that helped repair his own cells—and also reduce his insulin resistance, another way in which exercise counteracts diabetes.

But my encounter with Todd Becker did get me thinking more about the topic of stress in general. And as I delved into the literature about stress and hormesis, I began to realize that much of what we think we know about stress is, in a nutshell, completely wrong.

Stress is a bit of a catchall term: We use it to describe how we feel when we're under pressure at work, driving on Los Angeles freeways, or churning toward a book deadline; we also "stress out" over getting to the airport on time, vacationing with in-laws, or anything to do with raising teenagers. This psychological stress can give rise to biological stress: One way is by triggering the release of stress hormones like cortisol, which helped our Paleolithic ancestors engage their fight-or-flight survival response and store more calories as energy for long, hungry journeys into exile. In the modern world, alas, cortisol merely makes people in office jobs get fat. Stress also appears to shrink our telomeres, those thingymabobs that protect our chromosomes. So does major depression. Another very interesting study found that people who were lonely, which is one of the most intense forms of psychological stress, had their genes for inflammation activated at a much higher rate than those who felt more socially connected. But some stress is also good: One recent study found that women who had undergone stressful or difficult pregnancies appeared to be protected from some kinds of breast cancer.

The most common kind of biological stress is one that you have probably also heard about: oxidative stress, which is caused by free radicals. And if you know anything about oxidative stress, you know that it is bad, but that these free radicals can be—must be—combated by taking antioxidants, either as supplements or in our food. Which is why, when you go to the grocery store, nearly everything from fruit juice to breakfast cereal to dog food brags about its antioxidant content on the package. Antioxidants, most of us believe, somehow "soak up" or counteract the effects of free radicals. They're a good thing, as everyone knows. Except they're not.

The discovery of oxidative stress was one of the unsung dividends of the race to build the atomic bomb. During research for the Manhattan Project, things sometimes got a little sloppy, safety-wise, and many people were accidentally exposed to large doses of radiation. When that happened, these poor radiation victims seemed to grow old extremely rapidly: Their hair fell out, their skin wrinkled, and they developed various kinds of cancers in short order.

Scientists pretty quickly found that ionizing radiation created large amounts of free radicals, which are oxygen molecules with one free electron. The unpaired electron makes them chemically angry, and they rampage around, looking for nice innocent molecules to react with and corrupt. This chemical bullying results in *oxidation*, which is also why exposed metal rust and cut apples turn brown; this "browning" is also happening inside our bodies.

(And it's different from the "browning" that takes place in the Maillard reaction.) Free radicals also damage cellular DNA, leading to cancer (among other things). The atomic scientists figured out they could help victims of radiation poisoning by feeding them "radioprotective" compounds, which basically scooped up all the excess free radicals. These were the first antioxidants. And they worked, sort of.

But free radicals were not thought to have any link to aging until one morning in November 1954, when a young scientist named Denham Harman was sitting around his Berkeley lab, not doing much work. Back then, scientists did not have to spend every waking moment writing grant applications, like they do now, so they could spend time in the very productive enterprise of just sitting and thinking about stuff. Harman had been coming into the office like this for four months, just mulling over the single problem that consumed him: aging. Since every animal grew old and died, Harman believed, "there had to be some common, basic cause that was killing everything," he told an interviewer decades later. That morning, the answer dawned on him in one blinding moment: "Free radicals flashed through my mind," he later recalled.

Harman was ultimately able to prove that all oxygen-consuming life-forms produce free radicals, negatively charged oxygen molecules that chemists now call reactive oxygen species, or ROS, in their mitochondria. These molecules, he believed, caused the DNA damage and other cellular harm that drove the aging process—and also caused cancer. In particular, he believed, oxidized LDL cholesterol ("bad" cholesterol) appeared to be responsible for forming arterial plaques. They also damage proteins in our cells. But because ROS are an unstoppable, inevitable consequence of aerobic respiration, this oxidative damage is the price we pay for living in an oxygen-rich atmosphere. Or, to put it another way: Breathing kills.

Which is a depressing thought. But Harman had a solution: antioxidants. In one famous experiment, he fed mice food that was laced with the preservative BHT, which also happens to be a strong antioxidant, and they lived 45 percent longer. He did this again and again with other compounds, including vitamins C and E. His theory got a big boost from Nobel laureate Linus Pauling, who embraced vitamin C—one of the most powerful natural antioxidants—as a cure to all ailments, from the common cold to cancer.

The free radical theory of aging was the perfect 1960s theory, pitting "free radicals" against the forces of order, in the form of antioxidants. It was seductive in its simplicity: Aging was thus reduced to a chemical reaction that could be slowed, perhaps even stopped, simply by consuming a few pills. Harman himself took large amounts of vitamins C and E, the two most common antioxidants; he also ran two miles every single day, well into his eighties. He passed away in November 2014 in a nursing home in Nebraska, but his theory has taken over the world. More recently, health-foodies have embraced more exotic antioxidants, from beta-carotene to various phytochemicals found in blueberries, pomegranates, and red wine grapes, to name a few. More than 50 percent of the American public consciously takes some sort of antioxidant supplement, according to some estimates. Dr. Oz celebrates them daily on his show. The free radical theory, it's safe to say, has been almost universally accepted by the public.

But there was one nagging problem, gerontologists began to

realize in the 1990s: Antioxidants didn't really seem to extend lifespan in lab animals. Even Harman was slightly troubled by the fact that, while he could increase *average* lifespan, he was unable to extend the animals' *maximum* lifespan as he had in his early experiments. It was not clear that his antioxidant compounds were affecting the actual aging process. If they really were slowing down aging, by absorbing free radicals, then the longest-lived mice should have lived longer. (Linus Pauling, on the other hand, consumed massive doses of vitamin C, and lived to be ninety-three.)

Others had a difficult time replicating Harman's results. And in human studies—particularly the double-blind, randomized clinical trials that are the gold standard for medical science—antioxidant supplements have had a mixed track record at best. A massive *JAMA* review of sixty-eight clinical trials of antioxidants, covering a total of more than 230,000 subjects, found a huge disparity in results: A handful of studies showed that antioxidants reduced mortality risk, but overall the majority of well-run studies found that people who took vitamin A, vitamin E, and beta-carotene actually seemed to *increase* their risk of death. Beta-carotene in particular, once the darling of nutrients, was strongly associated with increased cancer risk—especially in smokers, for some strange reason. The data on vitamin C was less conclusive, and there was some indication that selenium might be slightly beneficial.

But, still: WTF? The antioxidant theory had seemed so simple and elegant. And now it appeared to be wrong. And not merely wrong, but antioxidants actually appeared to have killed thousands of people in large clinical trials. If free radicals really did cause cellular damage, and thence aging, then why didn't antioxidants seem to help?

In 2009, a maverick German scientist named Michael Ristow

shed some light on the issue with a simple but subversive experiment: His team recruited forty young people to a regular exercise program where they worked out for more than ninety minutes, five days a week. Half the subjects were given an antioxidant supplement with high doses of vitamin C and vitamin E, while the other half unknowingly took a placebo pill.

We know that exercise sharply increases oxidative stress, at least in the short term. For a long time, scientists puzzled over this fact, concluding that exercise is healthy *despite* the higher levels of ROS that it produces. And indeed, it's possible to exercise to excess, to the point where we cause actual damage; if we go lift weights one day after doing nothing for a month, we'll be sore afterward thanks in part to oxidative damage, or so it was thought. So for decades, athletes have been taking antioxidant supplements, thinking they would mitigate the soreness caused by their training.

Ristow's findings turned the oxidative stress theory upside down. He subjected his young, hopefully well-paid volunteers to excruciating muscle biopsies both before and after the training period. Both groups showed evidence of oxidative stress in their muscles after training, as expected, but he found that the subjects who had taken the supplements—harmless-seeming vitamins—had actually benefited far *less* from their exercise program than the ones on the placebo. If anything, the antioxidants seemed to have wiped out the benefits of training. All of which led Ristow to suggest, in a subsequent letter to the editor, that antioxidant supplements are "worse than useless."

He thinks so, and here's why: Normally, our bodies produce their own antioxidants, powerful enzymes with superheroesque names like superoxide dismutase and catalase, which soak up the excess free radicals that exercise produces. The supplements seemed to be blocking these enzymes. "If you take antioxidants, you preclude your own antioxidant systems from being activated," Ristow explains. "Not only antioxidants, but other repair enzymes."

In other words, by taking supplements we allow our native antioxidant defenses to grow weak and lazy, leaving us more vulnerable to damage from ROS. Adding a little bit of stress, like exercise, helps keep our own antioxidant defenses in tune. We adapt to the stress, and come out stronger (not to mention live longer). "This is why the benefits of exercise last so much longer than the exercise itself," he explains. The same goes for caloric restriction, he feels.

The supplements were blunting this hormesis response. The good news here is that we'll all be saving a lot of money in the vitamin aisle at Whole Foods—just like Todd Becker, the Hormesis Guy, who disdains pretty much all supplements, even things like fish oil and omega-3 fatty acids. As for things like pomegranates and blueberries, which have become celebrated as "superfoods" because of their antioxidant properties, Ristow says, "They are healthy *despite* the fact that they also happen to contain antioxidants."

Ristow's work also pointed to a completely new understanding of oxidative stress itself: Not only was it not harmful, but he suspected it was beneficial, and perhaps even essential to life. In studies, he and others had found that increased levels of oxidative stress actually *lengthened* lifespan in worms. Even spritzing them with low doses of the herbicide paraquat, which incites major oxidative stress because it's a poison, ended up making the little buggers live longer. Arsenic did the same. (Again: Don't try this at home.) Others had observed similar results: Even mice that had had their antioxidant enzymes completely whacked out of them (via genetic manipulation) suffered no reduction in lifespan or health. It got stranger. In other experiments, Ristow found that antioxidants also wiped out the benefits of calorically restricting his worms, which would normally extend their lifespan.

In English: Antioxidants appeared to be irrelevant to aging. And it didn't matter whether they are produced by our own bodies, or swallowed in pill form. Ristow went so far as to suggest that free radicals are far from the dangerous toxins that Denham Harman imagined; in reality, they are essential signaling molecules, produced by the mitochondria specifically in order to trigger beneficial stress responses in other parts of the cell. Or to simplify, ROS are our intracellular Paul Reveres. Exercise and lack of food—which in evolutionary terms means we are hunting or undergoing a famine—actually shift our mitochondria into a different state altogether, which he terms mitohormesis, in which they produce more free radicals, which in turn trigger our bodies' own health-promoting responses to stress, like DNA repair, glucose scavenging, even killing potential cancer cells.

A little bit of stress, then, is good for us. But what about a *lot* of stress? To answer that, I traveled to Texas in search of the world's most stressed-out animal, and also one of the ugliest.

In a windowless basement room, somewhere beyond the outskirts of San Antonio, the strangest rodent I have ever seen is crawling up my arm. She is bigger than a mouse but smaller than a hamster, and really weird-looking: pale pink, with delicate wrinkly skin and two pairs of enormous beaver-teeth. I email a pic to my girlfriend, who replies that it resembles "a penis with fangs."

Quite.

But the two most interesting facts about this critter, a naked mole rat whom I've dubbed Queeny, are these: She is thirty years old, six times as old as the oldest laboratory mouse ever known. And, two, she is pregnant. Very, very pregnant. So pregnant that I can see the outlines of her next litter, little dark lumps bulging beneath her nearly translucent skin. So, in human terms, I am holding in my hand the equivalent of an eight-hundred-year-old pregnant woman. Naked mole rats, it seems, have conquered menopause.

That's not all that's fascinating about Queeny: She and her brothers and sisters (and parents and children and uncles and cousins, all of whom are living together in this sealed Habitrail-like enclosure) endure massive amounts of oxidative stress in their bodies, levels that would scorch the fur off virtually any other animal on the planet. Yet it doesn't seem to bother them one bit. "From a very young age they have high levels of oxidative damage," says my guide, Shelley Buffenstein, "and yet they live twenty-eight more years."

More than three decades ago, as an undergraduate, Buffenstein had helped collect Queeny's parents and about a hundred of their friends and relatives from a colony that was living beneath a dirt road in Tsavo National Park in Kenya. At the time, not much was known about the naked mole rat, beyond the fact that they are really strange animals. Because they live underground, they had rarely ever even been seen. Unlike most rodents, mole rats are "eusocial," meaning they live in colonies dominated by a single breeding female, like many kinds of ants or bees. They are the only mammal to live this way. And they rarely if ever see the sun, which explains their delicate pink skin, not to mention the fact that they are blind. They "see" with their whiskers, Buffenstein informs me, in her lilting Afrikaner accent.

Buffenstein grew up on a farm in what was then white-ruled Rhodesia, and international sanctions had forced her to pay her own way through school in Kenya. So she took a job as an undergraduate lab assistant for a biologist named Jenny Jarvis, who was the first to study the naked mole rat. Jarvis kept a few mole rats in cages in her office, and when she embarked on an expedition to study them in the wild, Buffenstein went along. The result was that Queeny's ancestors were uprooted from their cozy underground nest, and when Buffenstein eventually immigrated to America, she brought the mole rat colony along with her. Here in the basement of her lab, they live in a rather large and complex Habitrail kind of system. One chamber is reserved as the communal toilet, just as in their underground nest. The air is quite pungent.

Originally, Buffenstein set out to study the animals' endocrine biology—hormones, basically—but as the years passed she noticed something odd about her already-strange charges: They seemingly refused to die. "When they turned ten, I thought, *Wow, they live a long time!*" she says. "I said, *We should start studying aging and figure out how they live so long.*"

Easier said than done. Everything about the "nakeds" (as she called them) seemed like a piece to a different puzzle, starting with their off-the-charts levels of oxidative damage; their tiny little bodies carried around more "rust" than a Russian roller coaster. Yet they kept on ticking for decades, long after the oxidative stress theory would insist that they should have died.

Their survival is all the more puzzling to Buffenstein, who believes that these extreme levels of oxidation are actually a function of their life in captivity. In the wild, deep in their underground burrows, the nakeds survive on much less oxygen than we surface dwellers enjoy; they rarely if ever see the sun, or breathe surface air. So they have much less oxidative stress. "Imagine sharing a burrow two meters deep with three hundred of your closest friends," she says wryly.

But in their lab cages, the naked mole rats are exposed to much higher levels of oxygen than they experience in the wild, which should poison them. "And yet they tolerate it!" she marvels. "They live in captivity for thirty years with all this oxidative damage."

She believes the mole rats pretty much disprove the oxidative stress theory of aging, at least in its simplest form. More interesting to her, though, is the idea that they could be undergoing a kind of stress response—our friend hormesis—that keeps them alive for decades longer than most other rodents. She thinks this stress response helps condition them for the stress of living in the colony, where a social hierarchy is rigidly enforced. "The queen is a bully," she explains, and within the colony there are power struggles worthy of Shakespeare, even over whose job it is to tidy up the brood's "toilet." The queen, being the queen, gets to poop right in the doorway. And as she is the beneficiary of all the others' work and protection, it is perhaps not surprising that she lives an astonishingly long time.

But then, so do the lowly workers who can last twenty years themselves. The fact that they live underground, where it's hard for predators to pluck them out (although snakes and spoonbill ibises certainly try), has earned them the luxury of evolving really long lifespans. Their bodies are simply better designed for growing old—which means they are better designed to handle stress, which they do by "up-regulating" their cells' own internal protective mechanisms.

For example, the mole rats' proteasome, the cellular garbage disposal, operates at a much higher rate than that of a normal mouse, which in turn helps it get rid of all those oxidatively damaged proteins and cell components. So where most lab mice die from cancer, and more die with it, the mole rats are almost completely cancer-free; Buffenstein's team has yet to find even a tumor in a dead mole rat. Their cells just don't become cancerous. Even when the mole rats were slathered with a highly potent carcinogen called DMBA, which induces pretty much instant cancer in mice, they remained healthy; it might as well have been suntan lotion.

"We think the naked mole rat cells have better surveillance mechanisms to say, *Wait a second, there's something not kosher, there's a change in my genome, and I'm not proliferating with it*," she says.

The naked mole rats are unusual, but by no means unique in nature. There are other animals that tolerate extreme levels of oxidative damage without dying. One of these hardy critters, a kind of cave-dwelling salamander called the olm, is found only in caves in Slovenia and northern Italy; ghostly pale, blind, and no more than ten inches long, the olm lives nearly seventy years—which is an insanely long lifespan for a salamander. It lives nearly as long as the average Slovenian male, which is why it's locally nicknamed "the human fish."

Indeed, the olms and the naked mole rats reminded me in some ways of Nir Barzilai's centenarians, like Irving Kahn; and not only because they were so small and wrinkly. They are hardly the world's most robust creatures, yet they both possess an inner resilience, a quality that author Nicholas Taleb dubs *antifragility*, that enables them to live for an extremely long time. Irving Kahn had escaped the heart disease and cancer that claim more than half of aging Americans—and this despite the fact that he had smoked for decades—because he is antifragile. Like the mole rats, he must be uniquely resistant to oxidative damage, and his proteasome probably kicks butt, too. Like the naked mole rats, his cells go to the Jaguar mechanics, not the Jiffy Lube guys.

But there's also a key difference between naked mole rats and centenarians, one that is perhaps the most important of all. Human lifespan, as we've seen, tops out around 120. Nobody really knows how long a naked mole rat can live—or an olm, for that matter, although there are very few of them in captivity to begin with. With the mole rats, it's not even entirely clear that they age, at least not the way other living things do; their mortality rates do not climb with age, the way ours do (according to Gompertz's law, which we talked about earlier). Nor do the mole rats go through anything resembling menopause, or reproductive aging—as Queeny rather amply demonstrated, shortly after my visit, when she gave birth to a litter of more than a dozen wriggling pups. Do they age at all?

In an effort to find out, a team including Buffenstein worked to sequence the naked mole rat genome. They reported in *Nature* that in terms of gene expression—which genes are switched on, and which shut off, a crucial barometer of aging status—a twenty-year-old naked mole rat was essentially the same as a four-year-old, young adult naked mole rat. This is decidedly not the case in humans, where patterns of gene expression and DNA methylation are increasingly considered to be like biological "clocks" for aging. And it meant that, in essence, the naked mole rats were not aging—or were doing so very slowly. Very, *very* slowly.

More than any other mammal, even more so than forty-year-old bats or two-hundred-year-old bowhead whales, the naked mole rats had achieved what gerontologists called negligible senescence. That is, they hardly aged. Which must be nice.

But what do the naked mole rats—and the Slovenian cave salamanders—really tell us about our own aging? The answer may well be "nothing," because these animals are so singular. There's no one factor that we could turn into a drug, no one gene that we could hope to mimic, at least that scientists have found so far. As with much else in aging, the answer is mind-bogglingly complicated (or, as the scientists say, "multifactorial").

But there are surprising ways in which we can boost our own stress resistance, and even our resistance to cancer. One is exercise, which we know about. Cold water swimming, well, maybe that, too. But the last one is, yes, hunger—but not the long, slow, grinding, Biosphere-plate-licking kind of hunger. (Been there, done that.) This kind of hunger goes well with a cheesesteak and a beer.

Chapter 13

FAST FORWARD

The belly is an ungrateful wretch, it never remembers past favors, it always wants more tomorrow.

—Aleksandr Solzhenitsyn, *One Day in the Life of Ivan Denisovich*

Among those on hand, when the Biosphere was unsealed in September 1993, was a young graduate student from Italy named Valter Longo. Although he worked in Roy Walford's lab, Longo had never actually met his boss in person, because when he had arrived at UCLA in 1992, the scientist was already inside "the Bubble," maintaining contact with his lab by videoconference. As Longo watched Walford and his colleagues step through the portal and into the blinding Arizona sunshine, he was horrified.

"When people come out of jail, they look okay," he remembers. "These guys looked like hell. That was when I decided that maybe caloric restriction was not such a good idea."

Longo, who bears a passing resemblance to the actor Javier Bardem, had come to the United States intending to become a musician, not a scientist. His plan was to study jazz guitar at the University of North Texas, which has a world-famous music department. To pay for school, he joined the U.S. Army reserves, which seemed like a safe bet until Iraq invaded Kuwait in August of 1990. His tank unit was hours away from shipping out when Operation Desert Storm abruptly ended. Afterward, the school asked him to direct the marching band, a job he considered supremely uncool, so he changed majors to study the biology of aging instead, which is how he ended up with Walford at UCLA.

Longo had started out working with yeast, and early on, he made a breakthrough discovery: One long weekend, he went out of town and neglected to feed his yeast colony. When he returned, he expected to find them all dead from starvation, which would have been no big deal, since they were only yeast. When he got back, he discovered to his surprise that they were not only still alive, but thriving.

"Just as a joke," he says, he tried to do it again as a proper lab experiment: caloric restriction taken to its absurd extreme. He took a dish of yeast, which normally live in a kind of sugar syrup, and instead gave them only water. Again, the starved yeast lived longer. "A *lot* longer," says Longo. (Yes, even yeast get old and die.)

What started as a joke now piqued his interest: Why would this happen? And what did it say about diet and aging? Maybe the important thing about caloric restriction is not how many total calories a creature consumes in a day, he thought, but what happens to it when it is *not* eating.

Everyone who has ever skipped a meal knows that we feel different, somehow, when we are fasting—sometimes worse, and sometimes better. In this case, religion has been way ahead of the scientists: Many world faiths incorporate some form of short-term fasting, from the Muslim month of Ramadan to Christ's forty days in the desert. Scientists took longer to figure out that fasting may have health benefits as well as moral ones. In the 1940s, not long after Clive McCay starved his rats, a scientist named Frederick Hoelzel obtained similar life-extending results by simply feeding his lab animals every other day. But his 1946 paper on the subject received virtually no attention until decades later.

There was another, even more fascinating study conducted in the 1950s in a Spanish nursing home. The staff doctors divided the patients into two groups of sixty, randomly. One group was fed the usual nursing-home fare, while the other was fed on an alternating schedule: One day, they would eat half the normal ration; the next, they would be fed about 50 percent *more.* Over the ensuing three years, the doctors found striking differences between the two groups. The normally fed patients had spent nearly twice as many days in the hospital as the on-again/off-again patients, and more than twice as many of them had died: thirteen deaths versus six.

Yet these studies were basically ignored until a plastic surgeon from Mississippi named Jim Johnson stumbled upon them while searching around in an online database. Johnson had battled with his own weight for years, and he was looking for a diet technique that would help him lose it once and for all. "I'm a backsliding fat person," he admits. He also happens to be fluent in Spanish, and when he read the old nursing-home study (which was published only in Spanish), he got excited and wondered if short-term fasting or alternate feeding might have more wide-ranging effects on human health. But there was still next to no recent "literature" on it.

Johnson eventually found his way to an NIA scientist named

Mark Mattson, who had been looking at the effect of fasting on mice. In 2007, he persuaded Mattson to collaborate with him on a small study in actual humans. Johnson recruited a dozen overweight and obese volunteers, all of whom had some degree of asthma, a complaint with its roots in inflammation. The volunteers ate normally every other day, but on the days in between they subsisted on a meal-replacement shake that provided only 20 percent of their usual calories.

They lost weight, which wasn't that surprising, but their asthma symptoms also cleared up, perhaps because the fasting had reduced the level of inflammation in their bodies. Clearly, fasting was doing something good for these patients, beyond simply reducing their body fat. Studies of Muslims during Ramadan have found a similar effect, and even Muslim athletes seem to perform better during the month of daytime fasting. Speaking of athletes, the pro football running back Herschel Walker was famous for not eating at all before game days—a complete flouting of the conventional wisdom. It didn't seem to hold him back from winning the Heisman Trophy and playing in the NFL for fifteen years. Even into his late forties, Walker was competing as a mixed-martial-arts fighter, thrashing guys half his age. (He's since retired from that, which was probably wise.)

To the rest of us, the idea of going without food sounds torturous, something that only the religiously devoted might attempt. But Mattson points out that actually, our bodies are hardwired to survive without food. "If you look at evolutionary history, we didn't used to have three meals a day, plus snacks," Mattson points out. "Our ancestors, even pre-human ancestors, would have to go extended periods of time without food, so the individuals that survived were the ones that were able to cope with this situation."

Intrigued by these preliminary results, Mattson began pursuing it and found that not only do short periods without food improve physical health, in the same ways that caloric restriction does, but it also actually seems to be good for the brain. He found that mice (and later humans) that ate on an alternating schedule were found to have higher levels of brain-derived neurotrophic factor, or BDNF, which promotes the health and connectivity of neurons. Also produced when we exercise, BDNF helps preserve long-term memory and staves off degenerative conditions like Alzheimer's and Parkinson's.

"When you're hungry, your mind better be active and figuring out how to find food, how to compete and avoid hazards to get enough food to survive," Mattson says. In other words, when we're hungry we want to kill something. That's evolution at work. Unfortunately, evolution did not endow us with strong willpower when it comes to food; in fact, quite the opposite. So not many people have the discipline required to cut their food intake by 25 percent, day after day. (Hell, most of us can't even manage to floss regularly.) There's a reason why Luigi Fontana's caloric-restriction study had just a few dozen participants. Sure, they had great cholesterol numbers, but who would trade places with them?

Fasting, on the other hand, has a finish line; there is relief on the horizon, in a day or two. Fasting for brief periods appears to provide many of the same benefits of caloric restriction, and is more achievable than a lifelong commitment to austerity. "Only 10 percent of people can do caloric restriction," says Valter Longo. "With fasting, it's maybe 40 percent."

One reason may be that, as Mattson and others have shown,

intermittent fasting—or if you prefer, intermittent eating—has greater benefits than caloric restriction. Better yet... those benefits seem to be independent of how many calories you eat. In other words, you can eat just as much as you did before, you just can't keep eating all the time. Sounds easy, right?

The corollary to this is that there's no one "right way" to do intermittent fasting. Other researchers have found weight-loss and other benefits from all kinds of eating schedules: from every-other-day fasting (which sounds rough), to twice-a-week fasting, to simple meal skipping; so many books on fasting have come out in the last few years that it's safe to say intermittent fasting is trendy. But unlike many trendy diets, there's solid science behind it.

One researcher, Satchin Panda at the Scripps Research Institute in San Diego, put mice on an eight-hour limited feeding "window" and found that they did not gain any weight, even though they ate a high-fat diet. For humans, that translates into skipping breakfast each day; or better yet, dinner. Perish the thought. But I actually tried this one for a while, and once I got used to it (nice latte in the morning, then nothing until 1 p.m.), I kind of liked it; at least I could do it, and I did seem to lose weight and feel sharper in the morning.

As one woman who has practiced intermittent fasting for weight loss for years advised me, "I learned to embrace a little bit of hunger."

Which is a useful thing to tell oneself, next time you're stuck on a long plane flight with no decent food. *Embrace the hunger.* It's good stress, after all. We're evolutionarily hardwired for it. And as Longo eventually discovered, the benefits of hunger go all the way down to the cellular level.

* * *

Back in his lab, Longo was trying to figure out why his starved yeast were living longer, and what that might mean for us. Digging down into the molecular biology of it all—long story short—he ended up unlocking a series of metabolic pathways that appear to regulate longevity. At the deep cellular level, metabolism and longevity are so closely intertwined they are basically inseparable.

These metabolic pathways all radiate from an important cellular complex called TOR, which is perhaps best thought of as like the main circuit breaker in a large factory. When the breaker is switched on, the factory (that is, the cell) goes humming along, forging amino acids into the proteins that are the building blocks, messengers, and currency of life. It's busy and messy, like Christmas season in Santa's workshop. When the breaker is off, the cell goes into more of a maintenance mode, "recycling" old damaged proteins and, by cranking up autophagy, cleaning up the junk that accumulates in our cells over time—like January in Santa's shop.

In an influential paper published in *Science* in 2001, Longo found that blocking the TOR pathway caused his yeast to live three times longer. This led him to believe that many of the effects of caloric restriction come about because the lack of nutrients shuts down TOR—an effect observed not only in yeast, but also in more complex critters like worms, flies, and mice. (Like the sirtuins, TOR is "conserved," meaning it appears up and down the tree of life.)

Turning down TOR also inhibits many of the growth pathways that appear to be connected with aging. With the TOR circuit breaker switched off, protein manufacturing is shut down, and cells don't divide as rapidly, so the animal does not grow. Instead, its cells become "cleaner" and healthier. They also resist stress better, and use fuel more efficiently—and thus are less susceptible to damage. It's a classic example of beneficial stress response, or hormesis. And it makes sense evolutionarily: When food is scarce, there is no point in wasting energy on growth.

But after his experience in Roy Walford's lab, Longo wasn't so interested in dietary restriction, which he considers to be "gradual, chronic suffering." But just as chronic, long-term stress is bad, short-term and acute stress can be good. Temporary, limited fasting qualified as short-term stress—and if anything, it appeared to shut down TOR more completely than a partial cutback in calories. So its effects were more intense. "Fasting is much more powerful than calorie restriction," Longo says. "It's like the strongest cocktail of medicine."

But medicine for what?

That's where the story gets really interesting.

One day about ten years ago, Longo heard from a friend of his named Lizzia Raffaghello, who was a cancer researcher at Los

Angeles Children's Hospital. She had a young patient, a six- or seven-year-old Italian girl who suffered from a rare type of brain tumor called a neuroblastoma. She wondered whether Longo could help the girl somehow. He said he couldn't—he studied aging, not cancer—and the little girl passed away soon after.

Her death prompted Longo to do some soul searching about his choice of career. "Lizzia and I got into a lot of discussions about whether it was right to focus on extending the human lifespan when a seven-year-old can die of cancer and there is nothing we know how to do to help her," he recalls. He was not an MD; he studied yeast. But he realized that his yeast might actually have yielded an insight about the nature of cancer.

When he starved the yeast, they not only lived longer, but they became immensely resistant to stress of all kinds, like oxidative stress caused by free radicals, and exposure to toxins. Meanwhile, although tumor cells seemed invincible, he knew that they actually were not. The reason is that cancer cells must eat constantly, gorging themselves like André the Giant at a cruise-ship buffet. One way doctors locate tumors is by injecting them with glucose that carries a chemical tag. The tumors hog all the glucose, so they light up with the tag. Longo saw that this made them potentially vulnerable. Because the tumor cells were always eating, always growing, their TOR was turned up to eleven—which actually *reduced* their stress resistance. In the lab, he showed that subjecting cancer cells to added stress, by taking away their food, really did weaken them.

He proposed a radical experiment to Raffaghello and her colleagues: Take mice with cancer, and starve some of them for as long as they could stand it, and then blitz them with huge doses of chemotherapy drugs, which are (obviously) highly toxic to all cells. "I still remember when I presented the idea to one of her MD collaborators in Italy," he says. "He looked at me shaking his head, and thinking that was the dumbest idea he had ever heard."

But the results surprised everyone: In some of the experiments, all the pre-starved animals survived the chemo, while all the normally fed ones died. The short-term fasting appeared to have switched the animals' normal cells into a protected state, while the tumor cells remained more vulnerable to the chemotherapy agents. This "differential stress resistance," as they termed it, could make the drugs more effective by targeting them at the cancer cells themselves; the cancer cells would be unable to adapt, while the noncancerous cells were in a protected state because of the fasting. So they would suffer less collateral damage.

Trying it out in human patients was not easy. Chronic calorie restriction had long been known to protect mammals against cancer—as it had done for the monkeys—but the severe weight loss it entails would seem to rule it out for use by actual cancer patients, who were already fighting to keep their weight. The oncologists were skeptical, and many were not willing to subject their already-suffering patients to yet more suffering. The patients weren't thrilled at first, either. "Nobody wants to fast, especially people with cancer," he says. "They're like, *What? You're telling me not to eat?* It just seems weird to people."

Doctors resisted it, too. But Longo and an MD in his lab named Fernando Safdie ultimately found ten late-stage cancer patients who were willing to give it a try on a voluntary basis. They fasted for between two and five days (!) in conjunction with a cycle of chemo, and surprisingly, all ten reported less severe side effects from the treatment after fasting. In some patients, the chemo also appeared to be more effective. It was only a tiny pilot study, but the results were intriguing enough that there are now five larger clinical trials going on, trying short-term fasting in conjunction with chemotherapy in about one hundred patients each, at USC, the Mayo Clinic and in Leiden, the Netherlands, and other locations. Early results have been promising.

"The key mechanism, we think, is really what I call death by confusion," Longo explains. "The idea is that normal cells have evolved to understand all kinds of environments, and cancer cells have de-evolved in some sense: They're very good at doing a few things but are just generally bad at adapting to different environments, especially if they're extreme."

Cancer cells are dumb, then; and when we fast, our healthy cells get smarter, or at least more adaptable to stress. And we wouldn't know any of it, if tourists had not discovered Easter Island.

A remote outpost in the Pacific, two thousand miles west of Chile, Easter Island is of course famous for its huge, mysterious statues of giant heads. In the 1960s, as the Chilean government was preparing to expand the airport in order to bring in more tourists, a Canadian expedition visited the island to take soil and plant samples before the isolated ecosystem was disturbed by outsiders.

In one of the samples, the scientists found a unique bacterium called *Streptomyces hygroscopicus*, which sounds like something you might catch from a bus-station drinking fountain but is in fact fairly benign—to humans, at least. In the dark subworld of the soil, however, a chemical war is being waged between bacteria on one side, and fungi on the other. Penicillin, for example, is produced by mold to kill bacteria—hence its antibiotic properties. Bacteria fight back with their own poisons. The Canadian scientists, from the Ayerst drug company in Montreal, found that *Streptomyces hygroscopicus* secretes an especially intriguing fungus-fighting compound that they named rapamycin, after Easter Island's native name, Rapa Nui.

The Ayerst team initially saw rapamycin as a potential antifungal drug (think Dr. Scholl's athlete's foot spray), but in the process they discovered that it had even more powerful effects on the human immune system, damping down the body's response to invaders. Not only that, but there were signs that it could do other things as well. But Ayerst was not interested, and the company soon shut down its Montreal lab and fired most of the staff. The scientists were ordered to destroy all their current projects, but Suren Sehgal, the Indian-born researcher who had discovered rapamycin, was not about to do that. Instead, he smuggled his samples out of the lab and stashed them in his wife's freezer, where they languished for five years. Thanks to his persistence, rapamycin was eventually rapamycin was eventually approved by the FDA in 1999 as a drug to help prevent transplant patients from rejecting their new organs.

Which made it a useful but somewhat obscure drug. But rapamycin's impact ultimately went way beyond transplant patients, and led to a completely new understanding of cell biology. Researchers investigating its mechanism of action eventually discovered TOR, the key growth regulator of the cell—*TOR* actually stands for "target of rapamycin."

In other words, this strange chemical, produced by a microscopic organism that lives in the dirt on an island two thousand miles from land, just happens to unlock the master growth switch for nearly all forms of life on this planet. No big deal.

But that was only the beginning. In a major study published in 2009—the very same day the calorically restricted monkeys made the front page of the *New York Times*, in fact—a team of NIH-funded researchers found that rapamycin had significantly extended the lifespan of mice. This was huge news, maybe even bigger than the monkey study: No other drug had ever extended *maximum* lifespan in normal animals before—how long the oldest animals lived. (Resveratrol had only worked on fat mice.) And it confirmed what Longo's lab had observed a decade earlier, that shutting down TOR also seemed to slow down aging.

Not only that, but rapamycin had worked even though the mice were already middle-aged when they took it. The study had started late, because the team's pharmacologist had spent months trying to get the drug into mouse feed in a chemically stable way. By the time he figured it out, the animals were already nearly twenty months old, the mouse equivalent of about sixty human years—too old, according to conventional wisdom, for an anti-aging drug to have any effect. Yet still the stuff had increased both the average *and* maximum overall lifespan of the animals by 9 percent for males, and 14 percent for females. Which may not sound like a lot, but given the late start it was the equivalent of giving sixty-five-year-old humans an extra six to eight years of life, or a 52 percent boost in remaining life expectancy.

Steven Austad was one of the study authors, and he noticed that not only had the mice lived longer, but their tendons were more elastic—just like the long-lived, slowly aging possums he had studied decades ago on Sapelo Island. That was a pretty good sign that rapamycin was actually slowing aging in the mice—everywhere but in their testicles, which suffered from a mysterious degeneration.

So much for *that* wonder drug, you're probably thinking, but researchers seized on rapamycin, and before long, more positive evidence rolled in. Rapamycin appeared to reduce the incidence of cancer and, more interestingly, it seemed to slow the formation of senescent cells. Even more dramatic was the 2013 finding, by Simon Melov and other scientists at the Buck Institute, that rapamycin actually reversed cardiac aging in elderly mice. After three months of rapamycin treatment, their hearts and blood vessels were actually in better shape than when the study started. "Their heart function had improved over baseline, so it had actually gone backward, which was very, very impressive," says Melov.

The researchers also found that rapamycin reduced the level of inflammation in the mouse hearts—and that it was perhaps working on aging at a deeper level. "One of the big mysteries of aging is why do we get this pro-inflammatory response with age? Nobody really knows," says Melov. "We found that the heart is chronically inflamed in old animals, which as far as I know is novel. And rapamycin, lo and behold, reduced that inflammation."

It even improved the strength of the animals' bones. There seemed to be just about nothing that rapamycin couldn't fix. Here was a drug that appeared to delay many of the effects of aging—even when taken in middle age or later. And it was already approved by the FDA. Why not try it? Yet while several researchers I met admitted taking resveratrol, most notably David Sinclair, nobody copped to using rapamycin—with one exception.

There were maybe two or three people in the world who were not surprised that rapamycin extended lifespan, and one of them was a fascinating Russian émigré scientist named Mikhail Blagosklonny. Considered eccentric even by the standards of aging researchers, he is nevertheless respected as one of the more original thinkers in aging research.

Born and educated in St. Petersburg, where he received both an MD and a PhD in cancer research, he now works at the Roswell Park Cancer Institute in Buffalo, New York, perhaps the closest thing to an American Siberia. From his remote outpost, Blagosklonny had been arguing for years that rapamycin could be the magic bullet that aging researchers had been looking for. In a small but prescient paper published in 2006, Blagosklonny had predicted that rapamycin would likely extend lifespan in mammals. Three years later, the RapaMice proved him right. This was huge, in his view: There was already a drug on the market, safety-tested and approved for use in people, that actually seemed to slow aging. Moreover, its molecular target happened to be the most potent lifespan-extending pathway ever discovered. "It is an extraordinary luck that such a drug exists," he exulted.

But the really interesting part was his reasoning, which pointed to a completely new, counterintuitive understanding of how aging actually works. Going back to the mid-1900s, most scientists believed that aging was the result of damage that accumulated over decades, eventually leading to cellular dysfunction, which then caused age-related disease. Basically, our cells got dinged up until they no longer worked properly. That was the reason there was a Hayflick limit. Aging happens due to loss of cellular function.

But as Blagosklonny thought about cells in general, and cancer cells in particular, Blagosklonny began to realize that in fact, the reverse is true: Many of the bad things we associate with aging are in fact the product of excess cellular function. That is, our cells and our body systems work *too* well. Or too much. Cancer would be one obvious example: rather than dying out, cancer cells divide and grow ad infinitum—thanks to their hyperactivated TOR pathways. But it wasn't just cancer. He and others began to see many other aspects of aging as resulting not from a *loss* of cellular function, but from *runaway* cellular function. At some point after we stop growing, the engine that had powered our growth becomes an engine powering aging.

"We are programmed to function at a high level because it gives you a lot of advantages at the beginning of your life," Blagosklonny said in an interview via Skype (he dislikes travel, and avoids face-to-face interactions for the most part). "But after development is finished, it's like a car that is leaving the highway and going to the parking lot. If you run your car in a parking lot at seventy miles per hour, it will be damaged."

Valter Longo agrees, but with a slightly different twist: "The programs optimized for growth and development don't fail," he says, "they simply start contributing to the problem since there is no evolved purpose for them anymore." So aging isn't programmed, as August Weissmann had thought more than a century ago; it's more like a program gone wrong, or one that has outlived its purpose.

In the past, this did not matter much, because the vast majority of humans died before the age of fifty. Very few of us reached the "parking lot." Now we do, and hyperfunction becomes a problem. Hyperfunction is why 25 percent of women over seventy are diagnosed with breast cancer, versus just two percent of women under forty. It's why women in their fifties keep piling on fat to feed children they can no longer bear, and why the male prostate gland keeps growing through middle and older age, a leading cause of awkward TV ads about urination, and also of prostate cancer. Hyperfunction is also why we grow hair in our ears instead of on our heads where it belongs. And at the cellular level, the continuation of cell growth leads to cellular senescence, the toxification of our aged cells.

So Blagosklonny was not at all surprised that rapamycin had slowed aging in the mice. He expected it. He was itching with impatience—it should be tried in people right away, he argued. "It's not a question of should we do this, it's a question how," he said. Blagosklonny was so convinced by the mouse study that he had begun taking it himself, a move that would have made professor Brown-Séquard proud. "Immediately it worked—as a placebo," he joked. "After five minutes, I felt so good!"

Beyond that, his only evidence that it works is that his marathon times have actually improved in the five years since he began taking it. Yet he betrays no doubt.

"Some people ask me, is it dangerous to take rapamycin?" he said. "I want to write article with title, 'It's more dangerous to *not* take rapamycin, than to overeat, smoke, and drink and drive without belt—taken together.' "

The fact that you've read this far means, of course, that you are far too intelligent to take a powerful, possibly dangerous drug just because you read about it in a book written by an English major who has no business giving anyone medical advice. Also: While rapamycin has passed muster with the FDA, it is only approved for transplant patients who are, by any standard, already very sick. Whether or not it should be used by healthy people, as a preventive against aging, is another question entirely—one to which most experts answer in the negative. "I'd like to see fewer side effects of a drug like that before I'd start taking it," says Randy Strong, the University of Texas at San Antonio pharmacologist who figured out how to feed rapamycin to mice.

For one thing, there's the fact that it is approved as a powerful immune suppressor. The RapaMice had lived in a sterile environment, where they were exposed to few if any pathogens. Actual human beings, out in the germ-infested real world, could be increasing their risk of an infection. The second reason that taking rapamycin long-term might be a bad idea is because it seems to increase insulin resistance, which is a step down the road to diabetes, obviously a bad thing if your goal is to live to be one hundred.

There is some evidence that rapamycin might not be all that bad—in some circumstances, for example, it actually seems to *improve* immune function. Novartis, which markets a version of rapamycin as a cancer drug called Afinitor, has begun small pilot studies in older patients, with some positive results. A group of scientists at the University of Washington is beginning a novel clinical trial of rapamycin in pet dogs. But there are two more hurdles to its use as a drug to counter aging: First, the FDA does not (yet) consider aging a valid "indication," and second, such a drug would have to be completely, utterly safe, with almost zero risk. "It's got to be safer than aspirin," says Randy Strong.

Valter Longo wasn't much interested in trying rapamycin in people, much less taking it himself; he felt that it worked on pathways that are too central for normal cellular function to be used safely. But his work with the yeast, and TOR, and then cancer patients had led him to think more about the role of growth factors in promoting aging. His lab has identified several drugs that block the growth-hormone receptor, and that might be suitable for human trials. "It's the master switch," he says.

Still, whenever he gave a public talk, especially in Southern California, he would get asked the Growth Hormone

Question—sometimes even by doctors who prescribed HGH injections to their patients. He was convinced it was a bad idea, but people weren't getting the message, so in 2007, he got on a plane to Ecuador in search of definitive proof.

He eventually found himself in a car heading south on winding, ever-scarier mountain roads leading deep into the Andes Mountains. His destination was the remote, dirt-road village of San Vicente del Rio, where he was going to study a very unusual group of people. He had long suspected that higher levels of growth factors led to a shorter lifespan, but until now he'd had precious little evidence to work with. Then he had found out about Dr. Jaime Guevara-Aguirre, an Ecuadorean endocrinologist who had been studying a population of about a hundred individuals who lived in the mountains of southern Ecuador.

Called Laron little people, they possessed an extremely rare genetic mutation similar to that found in the dwarf mice (and in Chihuahuas, for that matter). Because of the mutation, their cells lacked growth-hormone receptors, meaning their bodies were essentially deaf to it, and as a result they only grew to be about four feet tall, or less. While the growth-hormone receptor mutation is only thought to affect about three hundred people worldwide, a cluster of more than one hundred of them lived in this remote part of Ecuador, scattered among several mountain villages. Locally, they were known as *viejitos*, "little old men," because they tended to be small and wrinkled, but inside, they were anything but old.

Guevara-Aguirre had been studying the Laron for more than twenty years; he was, in effect, their family physician. He had initially been curious about their small size, which was seen as a handicap. Indeed, perhaps as a result of their stature and their remote, rural environment, many of them drank heavily, and suffered from odd health problems as a result. They also seemed to get into a lot of fights, also likely due to their short stature. But despite this, another interesting pattern had emerged over time: None of them had ever died of cancer. And none had diabetes, despite the fact that one in five were obese, a far higher rate than average in Ecuador. Among the local population, 20 percent of their non-Laron relatives died of cancer, and at least 5 percent from diabetes.

Longo suspected that it was their lack of growth-hormone receptors that rendered them immune to both cancer and diabetes. In the lab, he had seen similar patterns in mice, worms, and even yeast: Reduced growth "signaling" to an animal's cells seemed to be strongly correlated with longer lifespans. The Laron did not necessarily live longer, as they tended to die from things like accidents and seizures (and drinking, and in fights), but what struck Longo was that for some reason few to none of them died of diseases of old age.

"They eat whatever the hell they want, they smoke, and they drink, and they still live pretty long," Longo told me. In fact, since his research began, they have begun to eat and drink even worse stuff than they used to, consuming fantastic quantities of soda and cake without guilt. "Now they think they are immune," he sighed. "They're getting cocky."

At least they were protected—unlike the Southern Californians who insisted on taking growth-hormone shots, flipping the "master switch" in the wrong direction. But so, too, was practically everyone else, because growth hormone and IGF can be triggered not only by expensive injections but by, say, a trip to McDonald's.

The millions of Americans who were gorging themselves on high-carb, fast-food diets weren't shooting themselves up à la Suzanne Somers or Dr. Life, but they might as well have been. When we take a big hit of sugar, like by drinking a Coke, our bodies respond by producing a surge of insulin, to help transport all that sugar to our cells. Not only do many of those calories then end up in our fat cells, à la Phil Bruno, but the insulin response, in turn, triggers good old IGF-1—insulin-like *growth* factor—which goes streaking directly at TOR, telling it to turn those calories into proteins, cells, growth. This is a good thing when we're young, but if we're not young anymore, then it makes bad things happen.

Longo's idea dovetails with much emerging research on nutrition, which is beginning to confirm that sugar and carbohydrates are far worse dietary villains than fat ever was, and that the true cause of rampant obesity, heart disease, and diabetes can be found in the mountain of sugar that most of us eat every year.

Longo even goes a step further, and suggests that high-*protein* diets may be just as bad as high-carb diets, and for the same reason: Excess protein intake also activates the growth hormone receptors, and TOR, the two main drivers of cellular aging. In an eighteen-year study published in March 2014 in the prestigious journal *Cell Metabolism*, Longo's team showed that middle-aged people who had eaten a diet high in dairy and meat were much more likely to die eventually of cancer—and that very high meat consumption is as risky, mortality-wise, as smoking. People who ate 10 percent protein or less—versus the 30 percent recommended in "healthy" diets like the Zone—lived longer, on average. And they were one-fourth as likely to die of cancer as the meat lovers.

Longo instead looks to different role models: the centenarians of Molochio, the village in Calabria in southern Italy where his parents happen to have been born. He has befriended one of them, a 109-year-old singer named Salvatore Caruso who still lives on his own—a pastoral version of Irving Kahn. Only he'd never touch a knish; he eats a far healthier diet, consisting of mostly green vegetables, some pasta, and a little bit of wine. Meat is an occasional luxury. Even Longo marvels at how little he consumes. The key to Caruso's diet, according to Longo, is its low carbohydrate content *and* its low protein content—intentionally or not, he thinks, it keeps his growth factors and TOR in check, which effectively slows the rate of aging. He and his cohort were trained to eat less, in effect, by history—war, poverty, and periodic famine.

"People his age, or eighty or ninety, they've been through bad periods and they understand," Longo says. "But that's what those bad periods did to them; they were fasting all the time."

One reason (voluntary) short-term fasting is so effective, Longo thinks, is because it can be uniquely tailored to each individual's physiology, not to mention hunger tolerance. But most of the time, you still get to eat more or less normally. Longo started doing it himself about ten years ago, he says, about the same time as he started working with the Laron, but his motivation was more personal—a particularly scary doctor's visit.

"You think you're pretty healthy, but then when you look more closely, you're not," says Longo. "My blood pressure already ten years ago was 140, my cholesterol was high; this is the story of half of the population of Europe, and maybe 80 percent of the U.S.: You find yourself in your thirties and forties, and you're starting to be a candidate for Lipitor, a candidate for hypertension medication, a candidate for cardiovascular medication—and next thing you know, you're a patient, you know! And what we're looking at is that you don't need to do any of these drugs. For 90 percent of the people, you can get rid of all of them, for life."

Longo models his own diet after the basics of Caruso's diet, plus what he's learned through his work. Most days he skips lunch completely, and at dinner he eats a low-protein, plant-based, vegan-ish diet designed to push down his IGF-1 levels (not just to keep himself rock-star thin at age forty-six). Once or twice a year, he'll put himself through a bare-bones fast for up to four days, taking in a bare minimum of nutrition, in order to "reset" his system. He believes this is the best option, based on mouse and human studies, and also because it happens to work for him.

Even his taste in automobiles reflects this on-off dichotomy: He drives to work in an ultra-efficient, all-electric Nissan Leaf, which entitles him to a primo parking space. But back home in Playa del Rey, there's a Ferrari sitting in the garage.

Chapter 14

WHO MOVED MY KEYS?

Let's spend the night together, wake up and live forever.

—Jamiroquai

On a spring evening in 2013, I finished up some interviews in Berkeley, then got in my rental car and joined the sclerotic flow of the evening rush on the 880, headed down the east side of the Bay toward San Jose. I was late, and as traffic slowed, I could feel the stress of freeway driving subtly accelerating my own aging process (along with the particulate pollution I was inhaling by the lungful). Eventually the traffic cleared, and I found my way to a small office complex in the town of Mountain View.

I had come for the monthly meeting of something called the Health Extension Salon, a loose collection of Bay Area people with an interest in aging research. I had been to many aging talks and conferences by then, typically serious affairs attended by scientists and a handful of self-taught amateur gerontologists, most well over sixty, listening to talks that quickly devolved into the dense alphabet soup that is modern molecular biology. When I walked into this room, though, I saw something that stunned me: *young people.*

The room was packed, with a standing-room crowd of about 150. The average age seemed to be well under forty; I could count the gray heads on the fingers of three hands. The wine was gone, my punishment for being late. This being the heart of Silicon Valley, the crowd had a definite tech start-up vibe: There were two guys in SpaceX jackets, and a quick saunter around the place confirmed that the host company was somehow involved in robotics. Later, promised the tousle-haired organizer, Joe Betts-Lacroix, there would be "Twister racing," an activity not normally pursued at aging-science meetings. But first, we were going to hear about stem cells and Siamese twins.

The featured speaker on this February night was a researcher from UCSF named Saul Villeda, and he fit right in. Dark-haired and stocky, he too was far younger than the typical lead scientist. He spoke more like a Southern California surfer than a jargon-spewing academic, spiking his sentences with the occasional "like." But he also proved to be a natural at explaining complex subjects to lay audiences, and when we met in his office a few days later, he revealed that he had honed this particular talent by discussing his work with his parents, who had emigrated from Guatemala with no education beyond the fifth grade.

When Saul was born in 1981, the family lived in East Los Angeles, where his father worked as a janitor, and his mother as a nurse's aide. They eventually saved enough money to realize the American dream and buy a house, but the closest place they could afford was way out in Lancaster, California, a small blue-collar town on the edge of the Mojave Desert. Saul did exceptionally well in school, with an aptitude for science, and he got into

UCLA on a scholarship. Next was graduate school at Stanford, where he ended up in the lab of a professor of neurology named Tony Wyss-Coray.

Along with another Stanford researcher named Thomas Rando, Wyss-Coray had helped to revive the old nineteenth-century technique of parabiosis—joining two animals together, so that they shared one circulatory system. Rando was interested in muscle, while Wyss-Coray had begun to look at the effect of old blood on mouse brains. Villeda stayed in his lab as long as he could.

He had recently established his own lab in UCSF's new Center of Regeneration Medicine and Stem Cell Research at the tender age of thirty-two, a fact that made him a bit of an outlier. Because of tight funding, most scientists nowadays consider themselves lucky to get their own lab before they turn forty or forty-five. A look at the history of science shows why this is a problem: Most major discoveries tend to be made by younger researchers, in their twenties and thirties, when they are still at the peak of creativity, still coming up with their good ideas. Einstein, for example, was just twenty-six when he wrote down his most famous equation, $E = mc^2$.

One reason for this is because, basically, younger scientists have younger brains: Their minds are plastic, creative, and rich with neurons that enable them to make the kinds of connections among observed and even obvious facts, and the intuitive leaps, that lead to great scientific discoveries. As we get older, even the smartest and most creative thinkers seem to stiffen and dry up, and run out of ideas.

And this, in a way, would be the subject of Saul Villeda's presentation.

It isn't pretty, what age does to our brains. The trouble stems from the unfortunate fact that, like heart muscle cells, neurons do not regenerate. (Not much, anyway.) So, just as a baseline, we typically lose about ten percent of our neurons over a lifetime. The problem is, we also lose more like one-quarter of our synapses, the *connections* between neurons that are crucial for every mental process. Not only that, but our neurons themselves become less connective, with fewer of the dendritic spines that let us form memories, thoughts, ideas.

This erosion happens very, very slowly, at first, but a recent *BMJ* study found that significant cognitive decline is already evident in many people by the time they reach their forties. We are not the only animal whose brains decline with age. Even fruit flies lose their memories. Scientists test for this by exposing the flies to plums, and giving them a mild electric shock. When they are given cherries, they get no shock. Eventually they learn that cherries are good, plums bad. Then, just a couple of weeks later—late middle age, for a fruit fly—they forget which is which. In a related story, I haven't been able to locate the remote control for my TiVo for months.

Fruit-fly brains have something else in common with ours: Over time, some of them tend to build up "plaques" made of cellular waste products in between their neurons, which cramps their style in a major way and sometimes kills them. This is the same kind of gunk that Dr. Alois Alzheimer, a Bavarian physician who ran the Frankfurt Asylum, observed inside the brain of his most unusual patient, when she died in 1906 at the age of fifty-six.

Her name was Auguste D., and she was the wife of a railway clerk who had become, literally, demented. She was confused and disoriented, had difficulty remembering things, and suffered from paranoia and hallucinations. She also accused her husband of having an affair with a neighbor, which may or may not have been the case.

She knew it was the eleventh month of the year, but believed that the year was 1800, not 1901. "She sits on the bed with a helpless expression," Alzheimer wrote in his intake notes. "At lunch she eats cauliflower and pork. Asked what she is eating, she answers *spinach*."

After she died, he figured out why. Her brain was a complete mess. Under the microscope, he could see that the space between her brain cells was filled with gummy plaques made of some mysterious substance. The neurons themselves were also disheveled, like tangled skeins of yarn. They were so striking that he sketched a few of them:

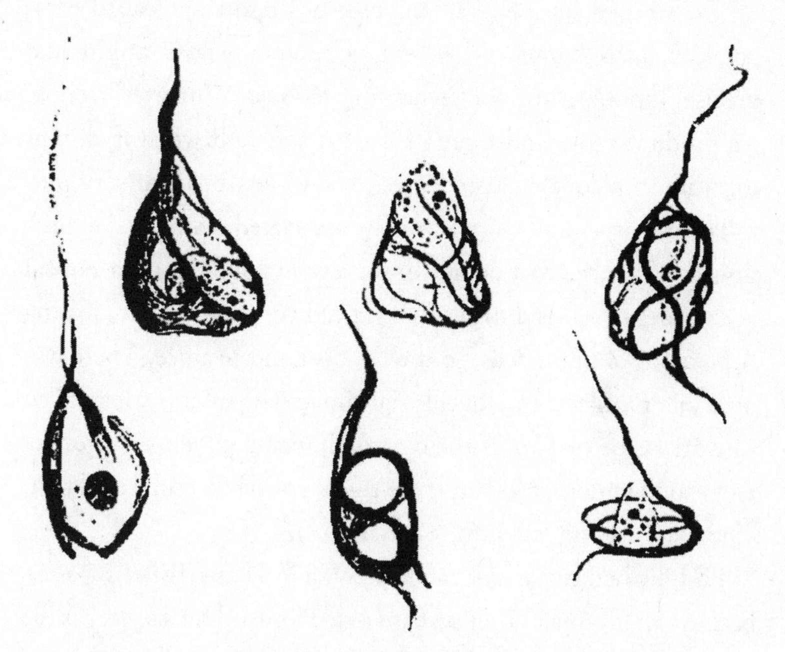

Credit: Bernard Becker Medical Library, Washington University School of Medicine

Alzheimer felt certain that these plaques and tangles had snarled Auguste D.'s thinking. A few years later, the syndrome was named Alzheimer's disease in an influential textbook. But it took until the early 1970s before it was recognized as the primary cause of what until then had simply been called senility. Already, Alzheimer's is listed by the CDC as the sixth-leading cause of death, but even that is misleading, because many patients actually end up dying from something else, such as an infection or heart failure. Something like 40 percent of Americans older than 84 are affected by Alzheimer's. By 2050, according to the Alzheimer's Association, the number of Americans living with the disease could more than triple, to sixteen million, and the costs to care for them will top $1 *trillion*.

The strange substance in Auguste D.'s brain was called beta-amyloid, also known as A-beta, a protein whose origin and precise function are somewhat mysterious. Whatever its job, we produce more and more of it with age, and when it clumps together in plaques it also proves toxic to neurons and very pro-inflammatory—and very strongly associated with Alzheimer's disease. Over the past decade or so, several major pharmaceutical companies developed drugs that proved very effective at clearing A-beta out of brain cells, in the lab dish and in mice. There was only one problem: In clinical trials in actual patients, they failed to work. One or two of them actually made patients' scores get *worse*, on memory tests similar to the ones I'd been subjected to in The Blast ("squid, cilantro, hacksaw…").

Eli Lilly had two major candidates fail in Phase III trials—a setback for the company, but perhaps a step forward for science, as the fiasco is forcing a new look at a hundred-year-old disease. More scientists are questioning the whole amyloid-causes-Alzheimer's theory itself; *Businessweek* magazine dubbed it a "drug graveyard." Out of more than two hundred potential Alzheimer's drugs that have been tested in clinical trials since 2002, only *one* has made it to market, and that one, called Aricept, is both insanely expensive and not all that effective (it delays the disease by about four months). Some researchers believe that another toxic protein called tau, which is also found in the brains of Alzheimer's patients, may be the actual culprit. Others believe the disease may start somewhere else entirely—and require a whole new way of treatment—or better yet, prevention.

What separates the Augustes of the world—the people who go into dementia in their fifties, for no obvious reason—from the Irving Kahns, who keep picking stock-market winners into their second century? How preventable is cognitive decline?

One surprising answer came from a study of 678 elderly nuns. Researchers from the University of Kentucky went back through convent archives and found autobiographies that 180 of the nuns had written when they were young and just entering the order. They analyzed the women's writing styles and found that the more nuanced and complex their sentences, and the richer their vocabulary, the less likely they were to have developed Alzheimer's or other forms of dementia. The more sophisticated writers also lived an average of seven years longer than those the researchers dubbed Listers, because their life stories amounted to little more than lists of names, dates, and places. On autopsy, it was found that the better writers' brains were also less gunked-up with amyloid than the Listers.

Another interesting, related observation came from The Blast. Autopsies of study subjects found that the brains of many cognitively "intact" patients were in fact loaded with amyloid plaques and tangles; they looked worse than the brains of some people who had actually been diagnosed with dementia. A British study found similar puzzling results: A third of "non-demented" subjects actually had massive amounts of junk in their brains—all the hallmarks of clinical Alzheimer's. Yet they had not developed any outward signs of the disease. Why?

One theory is that people with more education and more sophisticated, well-trained brains appear to develop what's called cognitive reserve, the way longtime athletes have built stronger, more stress-resistant cardiovascular systems. Education and learning develops more neuronal pathways and synaptic connections, so these people have in effect a buffer zone that protects them as degeneration begins to take place. It also gives them more tools with which to mask their cognitive impairment, consciously or not. Aging is hiding in their brains, just as it hides in our bodies. But it can't hide forever. When these smarty-pants patients eventually do develop the disease, or whatever it is, they tend to decline more quickly.

Use It or Lose It applies to our brains, too. It's similar to the old English farmer whom we met a few chapters ago, who retained his strong leg muscles well into his seventies, because he used them, every single day. A study of nearly two thousand elderly people published in June 2014 in *JAMA Neurology* found that those who had used their brains more from age forty onward were able to delay the onset of memory loss by more than ten years.

More research has shown that patients who resisted Alzheimer's often also resisted depression, which often goes hand in hand with brain aging. Those with a more "resilient" personality profile seemed better able to hold off cognitive decline, whether or not they had brain gunk. Similarly, the optimistic nuns also lived longer, by about seven years. By contrast, pessimists generally fared poorly, or at least their brains did. Depression eats away at our synaptic connections, truncating the size of the neural network, and eating into our cognitive reserve. Likewise, lack of sleep does much the same thing. Researchers are now realizing that sleep is absolutely crucial to brain health, especially in older adults; it gives brain cells a chance to clear out toxic or harmful metabolites, which would otherwise build up and further jam the network.

No wonder you feel dumb the day after an all-nighter. Even jet lag causes major disruption: In one study at the University of Virginia, scientists took a group of aged rats and advanced their light-dark cycle by six hours for a week, then by another six hours. Within four weeks, half of the rats were dead. (A shocker: Jet lag ages you.)

I don't know if this is good news or bad, but there are more things you can do that might help prevent Alzheimer's than there are drugs to treat it. A major 2011 study from UCSF found that seven basic middle-age risk factors, including diabetes, midlife obesity (defined as waist size of thirty-nine inches or more for men, thirty-six for women), midlife hypertension, smoking, depression, low educational level, and physical inactivity, could be responsible for as many as half of all Alzheimer's cases. More interestingly, a Buck Institute scientist named Dale Bredesen did a study where he actually addressed those kinds of risk factors in ten patients with full-blown Alzheimer's—especially diet, inactivity, and intellectual stimulation—and found that these simple lifestyle interventions had actually managed to achieve what no drug has yet been able to do: They reversed cognitive decline.

Exercise seems to be the key to keeping one's marbles: Yet another recent long-term study found that people who had been fitter at age twenty-five had stayed more cognitively "intact" at age fifty.

Mark Mattson of the NIH thinks there's a good evolutionary reason for this: Exercise sharpens memory so that we can better remember that we passed important things like sources of food, water, and building materials while we were out hunting. If you happened to walk past a promising hunting spot, or a fallen tree, or a spring, it was important to be able to find it again. So there may be a connection between my dad's maniacal bike riding—he cranked out twenty-five hundred miles between May and November 2013, more than I managed the entire year—and the fact that he's still mentally on his game.

Less-intense exercise also seems to have an effect: One well-done study found that merely walking twenty minutes a day was enough to slow or reverse the decline in cognition of patients who had already been diagnosed with Alzheimer's—something few drugs have been able to achieve. There's even an NIH-funded study going on to determine whether ballroom dancing has any helpful effect on brain function in seniors. It's probably safe to try it, without waiting for the results of this particular bit of taxpayer-sponsored research.

All of which suggests that, in part, the disease may have its origins in metabolism. Also, as Villeda points out, exercise itself also changes the "milieu"—that is, the chemical composition of our blood—in ways that seem to be favorable for the health of our neurons. When I would get stuck while writing this book, for example, I would stop working and go out for a one-hour bike ride; invariably, by the end of the ride the problem had been solved.

The only problem is that it's a temporary kind of effect. Still, a multitude of data is suggesting that the brain itself is far more plastic than we might have thought—and that even its aging may ultimately be reversible. So I may yet find that lost remote control.

Saul Villeda wasn't always convinced that old brain cells could be revived. In the lab with Tony Wyss-Coray, he had seen that the blood of Alzheimer's patients was markedly different from that of healthy older people, so the next question was whether those chemical changes in blood were somehow causing or promoting cognitive aging. "We were really asking, in terms of cognition, A, there's a blood-brain barrier, so does blood even have an influence on the brain? Number two, does old blood do something to the old brain?"

The question could only be answered via parabiosis. They joined several dozen pairs of mice together, just as Frederick Ludwig had done: old with old, young with young, old with young. After the critters had gotten to know each other, and had been swapping blood for a while, they would look for any change in the younger mouse brains. Several months later, they had their answer, summed up in the title of the resulting *Nature* paper: "The Aging Systemic Milieu Negatively Regulates Neurogenesis and Cognitive Function." In English, they found that young brains that were bathed in old blood (the "aging systemic milieu") worked poorly, with less protection and renewal of neurons than they should have had. Which is depressing: Old blood hurts your brain. But then Villeda started wondering: What about the other way around? What would young blood do to old brains?

The problem is that it's a bit difficult to tell what's going on in a mouse brain. For starters, you can't really do cognitive testing on a mouse that's sewed to another mouse. So Villeda tried another way: He would simply take blood plasma from young mice and inject it into old mice, and then run the old mice through a gauntlet of mental tests. Mice can't exactly take the SAT, so instead he placed them in a radial-arm water maze, a kind of Habitrail filled with milky water. Hidden just under the surface of the opaque liquid, in one part of the maze, there was a little platform that they could climb up onto to get out of the water. "These animals *hate* to get wet," he explained. "They'll do anything to avoid it."

Before any plasma injections, the mice went through a period of training in which they learned to find the safety platform in the water maze. Then, after a week or so, they were plopped back into the maze. Young animals could find the platform again almost immediately, while old animals would blunder around hopelessly, making as many as thirty errors before finally finding the nice, dry island. "It was kind of sad," Villeda said.

Then Villeda discovered something amazing: After the older mice had been injected with the young blood for a few weeks, they were suddenly able to find the platform again, on the first or second try. "Something is happening with the aging blood that is detrimental," he said. "There's something in young blood that we're losing."

After the mice were "sacrificed," he looked at their brains, especially the neurons from the hippocampus, the region where memories are formed. Under an electron microscope, younger neurons appear "fuzzy," because they have more dendritic spines, little branches sticking out from their neurons that help make connections with other dendrites (or larger branches). In old animals, the dendrites had many fewer spines, as if they had been pruned off by a zealous gardener. This made the neurons less connective, less able to link memories with thoughts and actions. But in the old mice that had been injected with young plasma, the neurons had become fuzzy again—which, obviously, had helped them remember and negotiate the water maze. Young blood had restored their old brains.

"We saw an actual reversal of aging-related decline," Villeda told me, still sounding amazed. "I always thought of aging as a final blow—once you get there, there's no coming back. I'm not sure that's the case anymore."

Now the big question was, *What* was it in young blood that produced this effect?

It wasn't only in the brain, either. Other studies had found that younger animals' blood also seemed to rejuvenate muscle and bone in older animals. And Villeda wasn't the only one looking for the answer. Across the country, at Harvard, another Stanford alumnus was searching for the precise factor that was responsible for this amazing rejuvenation effect. The race was on.

"It's not like looking for a needle in a haystack," said Amy Wagers, when we met in her office in Boston. "It's like looking for *hay* in a haystack. There are so many possible metabolites, or proteins, or factors, and any of them could be the one."

Her search had lasted ten years, since she had been part of the team that had helped revive the parabiosis technique in the early 2000s. As a postdoc, she had worked with Irving Weissman, the noted Stanford biologist who had first isolated human stem cells from blood; Weissman had collaborated with Tom Rando and his post doc, Irina Conboy, to see how aging blood affected muscle regeneration. In a groundbreaking 2005 paper published in *Nature*, they reported that young blood seemed to improve the ability of old mice to repair their injured muscles. Not only that, but their livers had also miraculously healed. Something in young blood was telling the old mice, on a cellular level, to act "young," too, and to regenerate and heal as successfully as they once had.

This meant that older cells still had the potential to thrive and regenerate, but the oldness of their blood was preventing this from happening. The implications of this were huge: It meant that we might retain the ability to regenerate various body tissues well into later life. The trick would be figuring out how to unlock that potential—by finding the factor, or factors, that might be responsible. The search would take a decade, and it isn't over.

Wagers decided to look for a possible rejuvenating factor:

whatever it was in young blood that seemed to turn back the clock on older cells. She was joined by a veteran cardiologist and stem-cell researcher named Richard T. Lee, who was also an old cycling friend of hers (and whom we met in chapter 6). Lee was getting tired of seeing his patients' hearts basically wear out with age as they lived into their eighties and beyond. Younger patients he could treat with statins and blood-pressure medication, and procedures such as stents and valves. But there seemed to be nothing he could do to solve these very old patients' problems. More and more of them suffered from something called diastolic heart failure, where the heart muscle has thickened so much that the chamber fails to fill properly; so far, there is no known treatment.

"After twenty years of watching it," he told me, "I was like, *Naaah. I got nothin'*."

Then Wagers suggested they try parabiosis, to see if young blood revived aging hearts. Their preliminary results suggested that it did, so they decided to try to find the factor that might be responsible for that particular effect. Working with a company in Colorado called SomaLogic, they narrowed the search down to thirteen candidates, all of which were growth-related factors of one sort or another. After more analysis, one clear winner emerged: something called growth differentiation factor 11, or GDF11, that was plentiful in young mouse blood, but not in old mouse blood.

Even better, biotech companies already manufactured GDF11 for research purposes, so they could simply buy it and inject it in their mice, which they did. And lo, it seemed to do the same thing as parabiosis, only without the creepy surgery: Just adding back GDF11 for a few weeks returned the aged, thickened hearts of the old mice to their normal, youthful state. It had set back the clock, they reported in May 2013. This was particularly interesting, because it had been thought that the "young blood" effect mostly worked on stem cells. But heart muscle doesn't have very active stem cells, so there had to be something else going on. Next, they looked at the effect of GDF11 on aged mouse muscles, and found that it improved their condition, too (good-bye, sarcopenia). More surprisingly, it also seemed to help aged mouse brains, by improving the blood vessels around neural stem cells. The elderly mice had even regained their sense of smell, they found.

When they first got their results, Wagers's old friend Lee wrote her a short email: "This could be big."

She replied, "I know."

They were right: Both studies were published in May 2014, in the same issue of *Science,* and they made headlines. Along with a third collaborator, Lee Rubin, Wagers and Lee are now pursuing possible drug candidates that would somehow activate GDF11; they have taken out patents, and are working with a venture capital group to fund the research. The goal is to come up with a drug molecule that mimics the activity of GDF11, or stimulates its production in the body. (GDF11 itself is too bulky and difficult to inject every day.) So far, they're tight-lipped, but the potential is huge: a drug that would treat heart failure, muscle wasting *and* Alzheimer's, potentially.

"This one protein is talking to many different cell types in many different tissues," says Wagers. "That is fundamentally interesting because it tells you why there might be synchrony in the response of different tissues to age."

But she also acknowledges that GDF11 is hardly the end of the story; it's more like the beginning of a new chapter. Two other major parabiosis-related papers were published the same week as Wagers and Lee's: In *Nature Medicine,* Saul Villeda finally reported his results, that blood plasma from young mice seemed to restore youthful sprightliness to old neurons. Across the Bay, in Berkeley, a Russian scientist named Irina Conboy, who had also trained in Tom Rando's lab, reported an even more intriguing result the very same day (it was a big day for parabiosis): Old muscles seemed to be rejuvenated by dosing with oxytocin, the "trust" hormone that's associated with sex, love, nurturing, and childbirth—it's even released when you give somebody a hug. Plus, it's cheap and easy to get. No transfusion required.

Not everyone is waiting for FDA approval: I know of at least one private scientist who has tried oxytocin injections on himself, hoping for some sort of rejuvenation effect. (No word yet on whether it's worked.) And Saul Villeda and his mentor, Tony Wyss-Coray of Stanford, are planning a small clinical trial of their own. Rather than look for the "hay in the haystack," they plan to simply inject plasma donated from young people into late-stage Alzheimer's patients, looking for signs of the rejuvenation effect that Villeda had observed in his mice.

If it succeeds, one can imagine all sorts of scary scenarios, like Donald Trump paying impoverished college students for plasma transfusions, so one almost roots for Wagers's team to succeed with their parabiosis pill. And there are still many unanswered questions such as whether these "youth factors" might cause cancer.

In the meantime, I'm just going to go for more hugs.

Epilogue

THE DEATH OF DEATH

Millions long for immortality who don't know what to do with themselves on a rainy Sunday afternoon.

—Susan Ertz

I've waited a long time to give you the really bad news about aging, but now I'm afraid I have to inform you that you probably have herpes. Not only that, but there's an even better chance that your mother does, too (or did). Also your father.

Don't feel bad: The herpes I'm talking about isn't the kind that you get on spring break. Rather, it's a way in to one of the most overlooked, yet potentially deadly forms of aging: the aging of our immune system.

Now, about that disease: There are, of course, many different kinds of herpes viruses, such as chickenpox and shingles, in addition to the kissing kind. But there is another, far more common version that is present in at least half of all American adults, yet (normally) causes no symptoms whatsoever. Most people do not even know they have it.

It's called cytomegalovirus, which sounds like something out of a science fiction movie, but in fact CMV (as it's known) is one of the largest and most promiscuous viruses in the human body, with an enormous genome that lets it attack just about any kind of human cell. It is usually fairly benign, hanging out quietly inside us without causing symptoms (it sometimes provokes a mononucleosis-type illness, but that's fairly rare). But that is not the same thing as being harmless. "There's some evidence that this virus can do some good things for you when you're young, like putting the immune system on a higher state of alert," says Janko Nikolich-Zugich, a leading authority on immune aging at the University of Arizona, "but it takes a toll going further down the line."

The problem stems from the way the human immune system normally works—and how it ages. The job of our immune defenses is to protect the body from the invaders, the unwanted, the attackers. With each new infection, we generate T-cells specifically tuned to fight that particular infection, which are sent out like shock troops to battle the latest contagion. All those T-cells emanate from the thymus, a spongy organ that sits more or less in the middle of your chest. If you've eaten at a nice French restaurant, thymus was probably on the menu as sweetbreads (or *ris de veau*). Starting around age twenty, yes you guessed it, your sweetbreads begin to shrivel up and die. It's one of the first things to go, and it goes almost completely. Which is bizarre, for such a seemingly important organ, but that's aging for you; the most important stuff is also the most vulnerable.

Of course, this "involution" of the thymus is irreversible, and eventually the thymus stops responding to new infections (though we do retain pools of T-cells that remember the infections we have already had, a rather tidy example of hormesis). So while our youthful immune systems could handle most unfamiliar bugs that we might encounter, seemingly minor bugs can kill us when we're older, if we haven't seen them before. Case in point: my otherwise robust grandfather, who succumbed to what started out as a pretty basic urinary tract infection. Immune aging is why older people should get flu vaccines; respiratory infections are not really considered a disease of aging, but they kill more people than Alzheimer's, and they disproportionately affect the elderly.

The real reason CMV poses a problem, Janko explains, is because it ties up a lot of the immune system's bandwidth, like your teenager playing video games and hogging all the Internet capacity so you can't stream *Antiques Roadshow*. "A single bug can occupy half of your immunological system," he says. People infected with the virus live three to four years shorter, on average. Yet humans have coexisted with CMV for hundreds of thousands of years. "The coevolution of this virus with us is mind boggling," he says.

Like a good evolutionary partner, CMV had no intention of killing us, until we started living longer. Its effects are rarely felt when we are young, but over the long run, we now know, the cytomegalomonster begins to weaken us and make us vulnerable—not only to infection, but also to other aging-related diseases. Like a hostile occupying army, its presence may help ramp up the levels of inflammation found in older adults, which make us susceptible to disease. CMV is particularly strongly associated with preclinical cardiovascular disease, because it targets the endothelial cells in the lining of blood vessels, causing inflammation that leads to arterial plaques—where the virus is frequently found lurking.

In short, it's like a guerrilla war, which is something Nikolich-Zugich knows a thing or two about, having grown up in what was once communist-run Yugoslavia. After the fall of the Soviet Union, he watched the country falling apart and realized he would have to make his scientific career elsewhere. He fled to the United States to study at Tufts, and in Boston he lived with a Jewish family that his own father had helped to escape from the Nazis during World War II. He ultimately landed at the University of Arizona, and began studying the immune system. When I met him, he was chairing a discussion on possible ways that we might, one day, be able to eliminate the aging of the immune system, and presumably thus his job, at a stroke.

The occasion was the sixth biannual SENS meeting in Cambridge, England, a conference put on every two years by Aubrey de Grey, the bearded, beer-drinking self-styled prophet of immortality, since 2003. The gathering was focused on research that might help put de Grey's complicated anti-aging strategy into practice. SENS, remember, stands for "Strategies for Engineering Negligible Senescence," and it was born in an epiphany that de Grey experienced, where he realized that the only way to achieve true longevity extension is by altering fundamental human biology, so that we more resemble—on a cellular level—critters like the 500-year-old clam, or the 200-year-old whales, or the thirty-year-old naked mole-rats, who barely aged at all. Merely tinkering with metabolism, which was and still is the dominant research approach, would only get us a few years at best he feels. (Hungry years, too.)

In its early years, SENS conferences were populated largely by fringey characters, but it has steadily grown more well-regarded, attracting respected scientists like Nikolich-Zugich. In his session, we heard from not just one but two well-known researchers who were working on competing ways to grow new thymus glands in older bodies. Neither method had really worked too well, so far, but it was a start.

De Grey himself had evolved since his *60 Minutes* heyday; he was no longer just the bearded guy in the pub saying wild things to Morley Safer about living for a thousand years. Nor was he affiliated with Cambridge any longer; the university had dismissed him for creating the impression, advertent or not, that he was on the faculty there. That had liberated him in a way, but even more game-changing was the fact that he now had money to fund actual research. This was thanks to his mother, who had had the prescience to buy two town houses in the Chelsea neighborhood of London in the early 1960s. When she passed away in 2011, they were worth more than $16 million, all of which went to her only son. Aubrey had used some of the money to buy himself a house in the woods in Los Gatos, where he now lives part-time along with one of the two girlfriends he maintains relationships with, while still being married to his wife Adelaide. ("It's well known that I am polyamorous," he told me.)

He plowed the rest of the proceeds into the SENS (Strategies for Engineering Negligible Senescence) Foundation, which meant that he could put his ideas to the test in labs. This itself was a major step: One of the knocks against him was that he had never done the "wet work" of hard bench science. Now he had funding for five years' worth of experiments. The grow-a-new-thymus project was just one curious project that his mom's real estate investments had helped to fund and promote. There was also a particularly interesting presentation on how a Mexican lake salamander, called the axolotl, can regrow its severed limbs, right down to the correct number of toes. If only we could tap that kind of regenerative power, then not only would there be no death, but there would be no more Thanksgiving finger-slicings, either. Fun fact: Not surprisingly, axolotl are loaded with telomerase. (They are also reportedly now nearly extinct in the wild, alas; so much for longevity.)

All sciences need the cranks, the fringers, the people who seem nuts; only time will tell if Aubrey de Grey is the one who turns out to be right. Already, there are signs that it's not completely crazy. For one thing, he had subtly changed some of his emphasis, so that he was not only pursuing the cellular strategies he had first outlined in his manifestos; now, he was embracing all sorts of regenerative biotechnology, some of which is downright attainable. Case in point: the thymus project, which may not be as far-fetched as it seems.

In April 2014, European scientists succeeded in growing new, functioning thymus glands in old mice by restoring a genetic mechanism that normally shuts down with age. The research was funded by a huge, $9 million EU-funded research initiative called ThymiStem, that was aimed at figuring out ways to regenerate the thymus gland in patients with damaged immune systems. Its founders had cancer patients and chemotherapy survivors in mind, but such a treatment would have an obvious and broader application in older adults. And it had crossed a major hurdle very quickly. So the idea that we might regrow or replace our crapped-out, old organs in some way—during the lifetimes of people who are reading this book—is not completely crazy.

In the pub at night, the beer and the conversation flowed freely. Few of the attendees seemed to doubt that aging and death would eventually be defeated, and that technology would solve all our other problems as well. To argue otherwise was to risk being labeled a "deathist." There was even a small protest against the dearth of funding for longevity research, held at a nearby pub (of course).

We already live in a world where biotechnology can bring back our lost, loved pets, sort of: At dinner one night in the vaulted Queen's College dining hall, the Silicon Valley finance guy next to me showed me photos of his cloned dog. I was envious. Across from us were two guys in their late twenties. They were roommates, sharing a flat in London as they worked on developing an app, the twenty-first-century version of writing the Great Novel.

"Why are you guys here?" I asked. "You seem a bit young to be worried about aging."

They looked at me like I was nuts. "Because we want to stay like this!" one said finally. "Wouldn't you?"

It was hard to argue the point. Who wouldn't want to stop the clock at, say, thirty?

But most of the conference presentations, on the other hand, made clear that death is a long way from surrendering anytime soon. Hell, we can't even grow a decent thymus gland yet—a small, simple organ with only three or four different types of cells. (None of the regenerated thymuses grown to date function particularly well.) How will we keep the whole human body from dying? Until then, I'm doing everything I can to keep myself from falling apart—although I'll stop short of signing up for cryonic preservation. I'm not sure I want to come back to life in the shape I was in at the exact moment when I died.

But the truth is that there is no such thing as a "cure" for aging, much less a "secret" (as I had to explain to practically every single one of my friends while working on this book). Aging science has pieced together the edges of the puzzle, but to paraphrase Philip Roth, we still don't know whether the picture in the frame depicts a battle or a massacre. The secret to aging? *Use it or lose it* may be the best we can do for now.

The morning after the conference ended, my head still a bit foggy from the pub, I fought my way onto a mobbed train and grabbed the only remaining seat, which happened to be next to one of my fellow conference attendees and pub buddies, a man named Sundeep Dhillon, who was rather an impressive character himself. In his twenties, he had become the youngest person to complete the Seven Summits, the highest peak on each continent, including Mount Everest. In 1996, the disastrous year chronicled in Jon Krakauer's *Into Thin Air*, he had turned back four hundred meters from the summit of Everest, and he had buried a deceased fellow climber on the way down. More recently, he had served as a combat doctor in the British Army in Iraq and Afghanistan. For fun, he had run the Marathon des Sables, 140 miles across the Sahara. In London, he worked as an ER doctor and also with a health-tech start-up.

He had spent a fair amount of time on the edge between life and death, in other words. I asked what he thought of the conference, and he immediately brought up the final presentation, which had dealt with the effects of aging research on population growth. SENS had assigned a Colorado demographer named Randall Kuhn to game out various longevity scenarios and their likely effect on world population—a prime objection of those opposed to longevity research. First the good news: If we're just talking about people living a bit longer, say to 100 or 120 or even 150, then world population won't grow all that much—which stands to reason, since no matter how healthy they are, hundred-year-olds probably aren't going to be having more children.

But the goal of SENS is not just to live a little bit longer, a little bit healthier; the goal is to live a really long time, in perfect, youthful health. Best-case scenario, that means women would not undergo menopause until much, much later—if at all. So a typical female could have four or five children, spread out over a much longer lifetime, instead of the current global average of two-ish. And that bumps the population curve up enormously.

According to Kuhn, world population is already headed to a likely plateau of ten billion. Delaying aging a little bit, like ten or twenty years, adds another couple of billion. But if SENS's negligible-senescence strategy is only mildly successful, Kuhn predicted, world population swells to seventeen billion by 2080. And if we add fertility into the equation, and people are able to live twice as long *and keep having children*, then world population hits one hundred billion by 2170. It was a sobering thought.

Then there were the economic aspects. Would all these new old people bleed the world dry? According to Kuhn, mega-longevity would actually be a net economic plus, as long as the retirement age was extended—say, to age 110. People would be far more productive, per capita. Unfortunately, due to the increased demand, food and energy prices would go "through the roof," he said, with oil hitting $1,000 a barrel.

"What about feeding all these people?" someone asked. Answer: If we have the technology to extend lifespan, we will probably also be able to grow or make enough food. Somehow. Perhaps we would live in the desert?

I wasn't so sure, and neither was Sundeep. He had seen war,

and he was not at all sure he would want to live on a planet with ten or twenty billion other people and a finite amount of arable land (not to mention fossil fuels). What would happen to the oceans? To the climate? Neither one of us shared the faith that technology would be able to solve this particular problem in an acceptable way. I started to think that, as dreadful as aging is, maybe immortality wasn't such a hot idea, either.

And there was a bigger irony at play, as well. The meeting had celebrated the rebels of science, the obscure seekers striving for great, world-changing breakthroughs (extreme longevity would certainly qualify). But then I thought of the physicist Max Planck's wry but true observation, "Science advances one funeral at a time."

Meaning that only when old scientists and their dogmas are retired does progress occur. If we figure out how to eliminate aging, then we'll eliminate scientists' funerals as well—actually, that will probably be one of the first things that happens. So how will science progress? What if Alexis Carrel had hung on for another fifty years? Would we still be celebrating his patently false dogma? Would Len Hayflick still be toiling in the basement of the Wistar Institute, grumbling that his cell cultures keep dying out?

More to the point: What if your boss would never, ever have to retire? Do you feel like being stuck in the same job until you're ninety-nine?

When we arrived in London, Sundeep guided me to the proper Tube line, and then we said our good-byes. Saturday is Football Day in England, and the subways and trains were thronged with soccer fans in full regalia, well-lubricated with the life-enhancing, life-extending beverage known as beer. Death and dying was very far from their minds, however, as it should be from all of ours, most of the time. More important was whether Chelsea would top Arsenal. Or whether my old friend John, whom I was going to visit, would slaughter me on the golf course (as usual).

And I thought about something I'd read from the philosopher Ernest Becker: "The idea of death, the fear of it, haunts the human animal like nothing else; it is a mainspring of human activity—activity largely designed to avoid the fatality of death, to overcome it by denying in some way that it is the final destiny for man." It makes us do things.

In other words, nearly everything we do is, on some level, motivated by the knowledge that someday we must die. It's why we write books, go to church, have children, take up pole vaulting at sixty, or quit our jobs and go find ourselves on the Pacific Crest Trail. Or whatever. "Death," as Steve Jobs famously and accurately said, "is life's change agent." It makes us do things.

I was surprised to come away from a weekend with Aubrey de Grey feeling slightly ambivalent about radical life extension. In my previous encounters with him, I'd been if not wholly convinced, at least optimistically curious. This time, I felt a bit more wishy-washy about it. My conversation with Sundeep kept replaying in my mind. What kind of world would we create?

Sure, I'd like to stay healthy for a long time, and enjoy my life the way I did when I was younger. That would be nice. But I want it much more for my surviving dog, Lizzy. I had always expected her to die well before Theo. She was a wild child, almost untrainable, always tearing off into the woods after deer and other critters. I figured that, sooner or later, she would be hit by a car, a common fate for hound dogs. It was okay: I had rescued her literally from death's door. Her owner had brought her in to my veterinarian's office (not Dr. Sane), to be put down for the high crime of chomping a Yorkie. I said I'd take her for a month, and that was twelve years ago. I've counted everything since then as bonus time.

We're all on bonus time, actually; think of our grandparents and great-grandparents' generation, who died in their fifties and sixties. After Theo, though, I expected the worst—ample research has shown that when one spouse dies, the other often follows shortly thereafter, and they were practically an old married couple, anyway. So I had relaxed my rules, letting Lizzy eat from the table and teaching her to beg for pizza crusts. At least she's eating, I told myself. *She'll be gone soon, anyway.*

It turns out she loves pizza crusts. And she is a very persistent and accomplished beggar. And she kept going. And going. And going.

For her thirteenth birthday, I went out and bought two juicy tenderloin steaks and grilled them up, one for her and one for me. Not long after, I took her in to Dr. Sane, the vet, for a checkup. "Thirteen!" he exclaimed, addressing her directly. "That's an accomplishment, Lizzy. Congratulations, girl."

She checked out fine, her blood work totally normal. She seemed to be embodying Tom Kirkwood's observation that the males drop dead while the females keep going. I wondered if I should get Nir Barzilai to start a centenarian-dog study, and look for her longevity genes, because she clearly had something protecting her from the cancer that kills most dogs by that age.

She was, to coin a phrase, no spring chicken. Her face had turned almost totally white, so she looked like a ghost. People began stopping us on the street again, like they used to when she and Theo were younger. "How *old* is she?" Complete strangers would come up to her and scratch her whitened muzzle, and sometimes even bend down to kiss her on the head, without a word to me, then walk off, wet-eyed. I loved her more than I ever had.

On a practical level, our life together slowed down considerably. It took us longer to negotiate the stairs up to our fifth-floor New York apartment, but she still managed. Some mornings she still trotted and pranced on our early walk, just like always; other days, she would just hobble a bit and do her business before turning for home. (She always dictated the route.) There was a scary episode, not long after she turned thirteen, when she basically lost her sense of balance, and staggered around like a drunk (neurological problem, cause unknown). I watched, helpless, until it eventually cleared up on its own.

All along, she got lots and lots of pizza crusts, particularly during the later stages of this project. I would have cloned her, if I could have afforded it. But as my dog-cloning Silicon Valley friend had explained to me, that night at dinner, the cloned dog wouldn't have been the same dog at all. If we could figure out how to make dogs live forever, I'd sign her right up.

As for me, I needed some help, too. Which brought me back to Nate Lebowitz's office, almost exactly a year after I'd first walked in. The waiting-room crowd was the same melting pot of New York–area immigrants and natives. My cholesterol numbers were not that much changed, either—down a few points on most things, up a good handful on HDL, the good stuff. That was okay. But Lebowitz was beaming. "Way to go!" he said. On my chart he'd written OUTSTANDING!

Why was he so happy?

A closer look revealed I had many fewer dangerous LDL-bomb-carrying mopeds zipping around in my arteries (the ApoB markers). Meanwhile, on the HDL side, I had somehow acquired more efficient arterial "sweepers," carrying cholesterol and other junk back out of the artery wall and to the liver for reuse.

Lebowitz wanted to know what I had done to make so much of an improvement: Had I taken Welchol, as he recommended? Nope. Fish oil? Sometimes. I'd also been more diligent about the bike riding, mostly trying to keep up with my dad. I had found a group of guys to ride with regularly, which helped a lot, though it would have helped more if our rides did not always end at a beer keg. Still, I had managed to give up burgers and French fries, two of my favorite foods. (Well, pretty much.)

Whatever I had done, it had worked: I'd also lost eight pounds, which is a pretty big deal. "Sometimes, just paying attention makes a difference," he said. True that. "You've healthy," was his parting verdict.

So is Lizzy, somehow. It is now the early fall of 2015, a perfect Indian Summer afternoon, and Lizzy is giving me her "hungry" bark as her rice and chicken boil slowly on the stove. I've been reduced to cooking meals for her, but anything to keep her eating. In barely a month, she'll mark her fifteenth birthday, putting her in the same longevity league as Irving Kahn. Better yet, she can still go up and down stairs, and has even been known to bound into my bed at 2 a.m. Not bad for a four-legged centenarian.

How did she get here? Perhaps she has dog centenarian genes, à la Irving Kahn, that have protected her (so far) from the cancer that killed her brother. Maybe it was the fact that, as I learned more about diet and aging, I started to feed her a bit less—always good, holistic foods, but never too much, sort of like the NIH/ Whole Foods monkeys. Or perhaps it was the fact that, nearly every day of her life, we've gone for a run, or a hike, or a good long walk. She used it, and didn't lose it.

Or maybe none of these things explains her longevity, and it was just pure dumb luck. I'd pay almost any price for a pill that would keep her going for three or four more years. But if it does come, it will be too late.

What does she feel like, inside that old body? I have no idea; I'm barely half her age, in dog years. Sometimes she romps like a puppy, flinging her squeaky toys around the room; other days, she's so stiff she can barely walk. Most of the time, she sleeps. She sleeps an awful lot, and once in a while I'll check to make sure she is still breathing.

In the morning, she wakes up slowly, stretching and yawning—but at a certain point, usually when I'm drinking coffee, she'll sidle up to me, tail gently wagging, her eyes telling me that it's time for our walk. I'll put down my coffee cup and get her leash, and out we'll go, down to the river or out on her favorite trail, breathing in the smells of the waking world together. Every day is a gift.

Appendix

THINGS THAT MIGHT WORK

Is there a magic bullet for aging? Not yet. But the following supplements and medications have been touted as "interventions" that might help ameliorate some aspects of the aging process. Some of them might even work. So, to satisfy your curiosity (and mine), I took a close look at the data on a handful of the most interesting possibilities.

Resveratrol

When David Sinclair's resveratrol-in-fat-mice study made the front page of the *New York Times*, in 2006, demand for resveratrol supplements hit the roof. The only problem: There were hardly any resveratrol supplements on the market. One of the few existing brands, called Longevinex, saw its orders skyrocket by 2,400-fold. While the hype has died down, resveratrol remains one of the biggest-selling "anti-aging" supplements on the market.

Certainly, resveratrol has amassed an impressive résumé of lab results, going back to before anyone ever heard of David Sinclair. When he "discovered" it, it was already well known for its ability to whack certain kinds of cancer cells, in experiments. Subsequently, it has been shown to improve liver function, reduce inflammation, prevent insulin resistance, and beat back some of the other effects of obesity. It also seems to improve cardiovascular function, such as in monkeys who were fed a brauts-and-funnel cake kind of diet. Yet despite the immense media hype it has attracted, for nearly a decade now, there have been only a handful of human clinical trials on resveratrol—versus more than 5,000 published papers on mice, worms, flies, monkeys, and yeast. Most of the human studies have been quite small, with just a dozen or two subjects, because nobody has seen fit to fund a large, well-run clinical trial of a non-patented supplement.

Worse, out of these small trials, very few have reported significant positive effects. Nir Barzilai found that it improved glucose tolerance slightly in older prediabetic adults. A couple of other small studies have shown slight beneficial effects on cardiac function. It appears to work best in animals and in humans that are obese or metabolically compromised; in them, it comes close to mimicking the effects of calorie restriction, without the restricted calories. But other studies have found no effect on insulin sensitivity, or cognition, or on blood pressure and other parameters, even in obese patients.

The reason for these disappointing results may have to do with the way humans metabolize the stuff. Even at very high doses, very little resveratrol actually makes it into our bloodstream, because our bodies think it is a poison, and it gets pretty much annihilated in the liver. Mice handle it differently, but in the human body, resveratrol does not last long; its half-life is around two and a half hours, or shorter than a baseball game. Another caution: Not all supplements labeled as resveratrol actually contain much resveratrol. Barzilai tested a dozen before he found one that was adequate for his study; a 2012 analysis of fourteen resveratrol supplements on the market found that five of them contained half or less the amount that was claimed on the label, and two had none at all.

This will not come as good news to whoever (besides my dad) buys the estimated $75 million worth of resveratrol supplements that are sold in the United States each year. But it will also not come as a surprise when you consider that the entire supplement market is left essentially unregulated by the federal government, thanks to the Orwellianly titled Dietary Supplement Health and Education Act of 1994—which neither educates nor promotes a science-based approach to health. Under the act, the FDA does not test or approve dietary supplements, so as a consumer, you are basically on your own. So you might want to start with the next entry on this list.

Alcohol/Red Wine

Here's a paradox: If drinking too much is bad for you, then why is it unhealthy to not drink? That seems to be the case: While excessive drinking is bad, an avalanche of studies has found that, overall, light to moderate drinkers are far better off than teetotalers, particularly in terms of cardiovascular health. A number of reasons have been suggested for this, but as we drill down into the data, we find that *what* you drink could be as important as how much. Oh, and in some cases, the more you drink, the better (up to a point, anyway). And the best thing to drink seems to be red wine.

Everyone knows red wine is good for you. It's thought to be at the heart of the French Paradox, whereby French people eat the equivalent of pork rinds and baloney ("paté") yet somehow don't get fat and unhealthy like Americans. Clearly, something in red wine must be good for you.

But resveratrol probably isn't the reason. The amount of resveratrol in a glass of red wine is minuscule, and as a large study of wine-swilling Italians showed, you don't get much—if any—resveratrol into your blood from even daily red-wine consumption. Yet still, red wine has been found to have beneficial effects on HDL (good) cholesterol, and blood pressure, and not just because it contains alcohol. In one extremely French experiment, which was actually done in Wisconsin, scientists infused Chateauneuf-du-Pape (1987) directly into the veins of dogs, so that they would not get drunk. They found that it reduced clotting and increased the elasticity of their circulatory systems.

Most curiously, and counterintuitively, red wine seems to protect against Alzheimer's. In a study done in Bordeaux (of course) red-wine drinkers were found to suffer from Alzheimer's less than half as often as their peers. And the Copenhagen City Heart Study found that the best-off study subjects, half as likely to die as the controls, were those who reported drinking between three and five glasses of the stuff. *Per day.*

Coffee

Another paradox: Not so very long ago, coffee was thought to be bad for you. This was undoubtedly because of the fact that it makes you feel good. Also, something about cancer. Yet without it, nothing would get done by anyone, anywhere. I'm on my third cup today, and it's nine o'clock at night. What to do?

Luckily, newer research is showing that, as usual, everybody in the past was completely wrong. It turns out many people in those old studies drank coffee and smoked at the same time, or they used to; hence the cancer findings. When scientists separated out the effects of smoking, they saw a much different picture. In 2012, a huge study in the *New England Journal of Medicine* reported that coffee drinkers tended to have significantly lower mortality risk than abstainers—but the really curious thing was that the more coffee people drank, the less they died. The association was linear, up to a point, which suggests that there may be some causation hidden in the correlation. In particular, it seems to lower the risk of Type 2 diabetes. Among those who drank four or five cups a day, overall mortality declined by 12 percent. Which means I need to go make another cup.

Curcumin

It's covered in the main text (see p. 85), but curcumin, a compound found in the spice turmeric, remains one of the most scientifically interesting compounds out there—and also one of the most frustrating. In the lab, it annihilates several different kinds of cancer cells, and some small studies have hinted at beneficial effects, such as on colorectal cancer, for example. But it faces the same "bioavailability" issues as resveratrol, only more so; the current thinking is that curcumin given with black pepper extract (as in, say, a nice Indian curry) might work better. You need to take huge doses of it just to see any show up in the blood.

At least one pharmaceutical company has tried to make it into a drug that the body can absorb, but as one pharma research chemist explained, if you alter the molecule to make it more absorbable, then curcumin becomes toxic.

"Life Extension Mix"

Sold by the Life Extension Foundation, which publishes *Life Extension* magazine and sells a vast range of supplements, Life Extension Mix sounds like the ultimate vitamin: The catalog says it is composed of "a broad array of fruit and vegetable extracts, as well as water- and fat-soluble vitamins, minerals, amino acids, and more." It contains many of the "healthy" nutrients found in vegetables such as broccoli and sweet potatoes, without the inconvenience of cooking and eating those vegetables. The ingredients list includes more than twenty separate miracle nutrients, including lycopene from tomatoes, olive juice extract, and of course blueberry extract. And every other healthy chemical that you've ever read about in the newspaper. But when it was put to the test by UC-Riverside scientist Stephen Spindler, who specializes in mouse lifespan studies—basically, feeding different things to large numbers of mice, and seeing if they live longer—Life Extension Mix flunked. Spindler's results showed that it and several other complex supplements "significantly decreased lifespan." Oh well.

Metformin

If you know someone with diabetes, or have it yourself, odds are pretty good that they (or you) are taking a drug called metformin, which is sold under brand names like Glucophage (also Fortamet,

Gluformin, and a half-dozen others). It's the most commonly prescribed diabetes medication, and costs just pennies per pill. What you don't know, and probably neither does your doctor, is that metformin is one of the most promising anti-aging drugs out there—and also one of the most mysterious.

Though it was discovered in the 1920s, as a derivative of French lilac, and its blood-sugar-lowering properties had been studied in the 1940s, how it exactly works remained unknown into the 1980s, when a young Israeli PhD student at Yale named Nir Barzilai wrote his thesis on metformin's likely mechanism of action. "Every few years since then," he jokes, "someone comes up with a new mechanism of action for metformin. So we don't know, okay? But what's happened with metformin that's amazing is there are lots of studies that show that people on metformin have less cardiovascular disease and cancer, and some are saying better cognitive function and things like that."

In general, metformin appears to reduce glucose production in the liver, which is more or less Grand Central Station for metabolism, and thus important for the aging process. And over the last few years, metformin has quietly racked up an impressive résumé of study results, many unrelated to diabetes. For example, it's been shown to kill cancer stem cells (in a dish, but still), and reduce the inflammatory response of cancer cells. More significantly, it is one of the few compounds that actually has been shown to extend lifespan in mice, who are notoriously hard to longevitize. (Is that a word?)

And it may even extend the lifespan of humans, as well. A major analysis of UK patient data found that diabetics who were taking metformin were actually outliving nondiabetics—which is astonishing, given that diabetics normally give up five to seven years of lifespan.

Barzilai and others believe metformin could be slowing the aging process itself. Later in 2015, the FDA could approve a large clinical trial of metformin in aging patients, which could lead to its approval as the first-ever legitimate anti-aging drug. The trial will not be looking at overall lifespan, but biomarkers, tipping points: Does it delay heart disease, reduce cancer risk, things like that?

"If somebody comes to me and says, I want to use a drug now, what would you say?" he says. "I would say metformin. I know how to use it, I know its safety, I know the studies."

Vitamin D

Another puzzle. Low vitamin D levels have been shown to be strongly associated with poor health and disease. Back in the old days, kids were force-fed cod-liver oil to beat rickets, a bone disease that stems from lack of vitamin D (which helps us metabolize calcium). Now we get vitamin D in milk—but not enough, apparently. Health problems related to lack of D persist, especially in Northern Europe and the United States and Canada; and especially among the elderly. One study found that 70 percent of whites in the United States had insufficient levels of vitamin D; so did 97 percent of blacks. The problem is that D is not found in many foods; hence the cod-liver oil.

Vitamin D is important for bone health, and also, overall physical performance. Data from the Women's Health Initiative study—the same folks who popped the estrogen-replacement balloon—found that supplementation with D and calcium (since the two are symbiotic) helped reduce the risk of fractures substantially. It also improves muscle strength, and newer data is suggesting that it might help put the brakes on the kind of cellular proliferation that leads to cancer. Lack of vitamin D is also associated with Parkinson's and Alzheimer's. Gordon Lithgow of the Buck Institute, who has been screening hundreds of compounds for anti-aging effects, thinks its effects might be much more far-reaching—in his lab, he says, vitamin D actually seems to slow down aging (in worms, but it's a start).

But other studies have found that supplementing alone does not help; one large meta-analysis said it did reduce overall mortality, while another was inconclusive. (Same old he-said, she-said health news story, in other words.) That's because it has to be activated in the body, argues Boston University scientist Michael Holick, who spent his thirty-five-year career researching vitamin D. And that job can only be done by the ultraviolet rays of the sun. In addition to supplementing with vitamin D_3, (at least 800 IUs per day, plus calcium), Holick advocates a little bit of sun exposure, a few times a week—sans sunscreen.

Aspirin and Ibuprofen

We've known for decades that a little bit of aspirin goes a long way toward preventing heart attacks and colorectal cancer. What we haven't really understood, yet, is why. As scientists increasingly recognize the importance of inflammation in aging and disease, the anti-inflammatories are looking better and better (aspirin and ibuprofen—but not acetaminophen, aka Tylenol, which has much greater health risks). They seem to help with cardiovascular health, which makes sense because inflammation is a necessary condition for forming atherosclerotic plaques. In an NIH study, aspirin also increased the lifespan of mice. And another study found that regular ibuprofen users were 44 percent likely (!) to get Alzheimer's disease.

Quercetin

A flavonol found in a diverse array of foods, from capers to carob to kale (see below), quercetin is a also popular supplement that, until recently, had very little good evidence to support its use. But a tantalizing 2015 study found that quercetin (along with a cancer drug called dasatanib) appeared to help remove senescent cells—those cellular bad actors that are thought to drive inflammation and other aspects of the aging process.

Kale

Why not? Try it with bacon.

Acknowledgments

This book would not have been possible without my parents, William Gifford, Jr., and Beverly Baker, who cultivated my love of reading, encouraged me to follow my heart, and also helped instill some fairly healthy habits. You are forgiven for not letting me drink Coke as a kid. My folks also both happen to be in magnificent health in their seventies, setting a high bar for the rest of us, so in a way they were the inspiration for this book.

It also would not have happened without the generosity of Nir Barzilai of Albert Einstein College of Medicine in the Bronx, who with his colleague Ana Maria Cuervo invited me to sit in on their graduate course in the biology of aging in the fall of 2011. That gave me, an English major, the strong grounding in the science that I needed, but even after the course ended Nir proved to be an indispensable Virgil in the aging world, introducing me to the people I needed to meet and providing helpful guidance.

I am grateful to the other busy scientists who gave me their time and let me tax their patience, including Rafael de Cabo, Mark Mattson, Luigi Ferrucci, and Felipe Sierra of the National Institute on Aging; David Sinclair, Amy Wagers, and Rich Lee at Harvard; Brian Kennedy, Judith Campisi, Gordon Lithgow, Simon Melov, and Pankaj Kapahi of the Buck Institute for Research on Aging in Marin County; Steven Austad, Veronica

Galvan, Randy Strong, Jim Nelson, and Rochelle Buffenstein (the naked mole rat lady) at the Barshop Center for Aging Research in San Antonio; Valter Longo, Tuck Finch, and Pinchas Cohen of USC; James Kirkland and Nathan LeBrasseur of the Mayo Clinic; Saul Villeda at UCSF; Donald Ingram of LSU; Mark Tarnopolsky at McMaster; and Jay Olshansky of the University of Illinois–Chicago. Also Leonard Hayflick, a great scientist and one-of-a-kind human being, who I feel fortunate to have met. And who can forget Aubrey de Grey, also sui generis.

As any journalist knows, some of the most helpful people you ever meet on a story are the passionate amateurs, the hyper-informed observers sitting just on the sidelines, the ones who know who and what you need to know. For me, that role was played by the inexhaustible Bill Vaughan and the omnipresent John Furber, among others. Eleanor Simonsick of the NIA also helped guide my research and my thinking, and while she did not want to be quoted, she didn't say I couldn't thank her. Michael Rae was also quite helpful. Dr. Nate Lebowitz helped me navigate the bewildering world of practical cardiology, and Charles Ducker helped me make sense of the biochemistry, no easy task. It was an honor to meet Irving Kahn. Finally, I doff my hat in admiration of Ron Gray, Howard Booth, and Jeanne Daprano, extraordinary athletes for any age.

My agents, Larry Weissman and Sascha Alper, provided encouragement, the occasional cattle-prodding, and a fantastic title. At Grand Central, Ben Greenberg took the chance that this would not be another sad, boring book about aging, and Maddie Caldwell, Yasmin Mathew, Liz Connor, and the rest of the hardworking GCP team kept things moving. Hard-charging GCP publicist

Linda Duggins would not take no for an answer. Thanks also to the talented Oliver Munday for a terrific cover and fun illustrations.

I shared the manuscript with trusted friends, including Jack Shafer, Weston Kosova, Alex Heard, Chris McDougall, David Howard, Jennifer Veser Besse, and Christine Hanna, who also offered her guest room in the Bay Area. Thanks also to my hosts Steve Rodrick and Jerry Hawke. Helpful commiseration was offered at various points by Jason Fagone, Carl Hoffman, Gabe Sherman, Brendan Koerner, Ben Wallace, Josh Dean, Max Potter and the amazing Leslie Morgan Steiner. I am indebted, literally, to editors who supported the project with aging-related assignments, including Glenda Bailey at *Harper's Bazaar*, Chris Keyes and Alex Heard at *Outside*, Laura Helmuth at *Slate*, and Michael Schaffer at the *New Republic*. Thanks also to Elizabeth Hummer for her support and patience.

And of course thanks to Lizzy the magical coonhound, who teaches me to love each day.

Notes & Sources

Prologue: The Elixir

p. xii, *"the most picturesque member of our faculty"*: William H. Taylor, "Old Days at the Old College," *The Old Dominion Journal of Medicine & Surgery* Vol. 17, no. 2 (August 1913). Taylor may or may not have been the student who discovered Brown-Séquard in his painted state.

p. xiii, *"may have been bipolar"*: Many details of Brown-Séquard's life come from the outstanding work of biographer Michael Aminoff, author of *Brown-Séquard, An Improbable Genius Who Transformed Medicine* (New York: Oxford University Press, 2010).

p. xiv, *"On June 1, 1889,"* Charles Edouard Brown-Séquard, "The Effects Produced on Man by Subcutaneous Injections of a Liquid Obtained From The Testicles of Animals," *Lancet*, July 20, 1889.

p. xvi, *"a young Elvis Presley"*: Brinkley's tale is recounted with amazing verve in Pope Brock's excellent *Charlatan: America's Most Dangerous Huckster, the Man Who Pursued Him, and the Age of Flim-Flam* (New York: Crown, 2008).

Chapter 1: Brothers

p. 7, *"its actual ingredients cost about $50"*: Buchanan had been hired to do the analysis by the UK newspaper *Daily Mail*. The resulting story appeared February 4, 2010.

p. 7, *"ten thousand Baby Boomers will celebrate their sixty-fifth birthdays"*: This figure was used by the U.S. Social Security Administration in its Annual Performance Plan for Fiscal Year 2012.

p. 8, *"than the death of old age"*: Michel de Montaigne, *Essays*, "To Study Philosophy Is to Learn to Die," in trans. by Charles Cotton, 1877.

Chapter 2: The Age of Aging

p. 15, *"The leading killer of Americans"*: Centers for Disease Control, "Leading Causes of Death, 1900-1998."

p. 16, *"a life expectancy of about seventy-seven years"*: Other agencies such as the CIA and the United Nations come up with slightly different numbers, but they tend to cluster around 76–78 years for men, 80–82 for women. Overall, the undisputed leader is Japan, where women can expect to live into their late 80s, statistically.

p. 16, *"their 105th birthdays"*: James Vaupel, personal communication.

p. 18, *"Doubts began to arise"*: Old Parr's story has been told in many places, but I first read about him in evolutionary biologist Steven Austad's excellent book, *Why We Age: What Science Is Discovering About the Body's Journey through Life* (New York: J. Wiley & Sons, 1997).

p. 18, *"an otherwise unremarkable Frenchwoman"*: Craig R. Whitney, "Jeanne Calment, World's Elder, Dies at 122," *New York Times,* August 5, 1997.

p. 20, *"back to eighteenth-century Sweden"*: This was, in fact, the origin of "statistics," which means "numbers in service of the state." The Swedish king demanded accurate population records, so he knew how many potential soldiers he had at his disposal, should he desire to teach those Norwegians a lesson once and for all.

p. 21, *"a steady rate of about 2.4 years per decade"*: Jim Oeppen and James W. Vaupel, "Broken Limits to Life Expectancy," *Science* 296(5570): 1029-1031.

p. 24, *"ninety-five might also be the new eighty"*: Kaare Christensen et al., "Physical and cognitive functioning of people older than 90 years: a comparison of two Danish cohorts born 10 years apart," *Lancet,* published online July 11, 2013.

p. 25, *"[lifespan may] begin to decline in some countries"*: S. Jay Olshansky et al., "A Potential Decline in Life Expectancy in the United States in the 21st Century," *New England Journal of Medicine* 352;11 (March 17, 2005) 1138-45. Olshansky later revisited his prediction in 2010, noting that in many areas of the U.S., life expectancies had already begun to decrease.

p. 27, *"Olshansky had confidently declared"*: S. J. Olshansky et al. (1990). "In search of Methuselah: estimating the upper limits to human longevity." *Science* 250(4981): 634-640.

p. 28, *"lower than in Guatemala"*: David A. Kindig and Erika R. Cheng, "Even As Mortality Fell In Most US Counties, Female Mortality Nonetheless Rose In 42.8 Percent Of Counties From 1992 To 2006," *Health Affairs* 32, no.3 (2013): 451-458.

p. 28, *"forty is the new sixty"*: Uri Ladabaum et al., "Obesity, abdominal obesity, physical activity, and caloric intake in US adults: 1988 to 2010," *American Journal of Medicine* 127(8):717-727 (August 2014). Similar data has been reported in numerous studies, particularly of the Baby Boom generation.

p. 28, *"large numbers of centenarians":* The "Blue Zones" identified by demographers include not only Okinawa but Sardinia, part of Costa Rica, and Loma Linda, California, home to large numbers of Seventh-Day Adventists. Dan Buettner, *The Blue Zones: 9 Lessons for Living Longer from the People Who've Lived the Longest.* (Washington, D.C.: National Geographic, 2012).

p. 30, *"lifespans of five thousand years or more":* De Grey has made a wide range of lifespan predictions in print; the five-thousand-year figure is floated in "Extrapolaholics Anonymous: Why Demographers' Rejections of a Huge Rise in Life Expectancy in This Century are Overconfident," *Annals of the New York Academy of Sciences* 1067 (2006): 83–93.

p. 32, *"His plan, which he calls SENS":* De Grey first outlined his idea before a group of aging scientists in 2000; the talk was eventually published as De Grey et al., "Time to talk SENS: Critiquing the Immutability of Human Aging," *Annals of the New York Academy of Sciences* 959 (2006):452–62. He explores and fleshes out his theories in *Ending Aging: The Rejuvenation Breakthroughs That Could Reverse Human Aging in Our Lifetime* (New York: St. Martin's Press, 2007), which is both highly detailed and fairly accessible.

p. 33, *"the idea of a Cambridge scientist who knows how to help us live forever":* Huber Warner et al., "Science fact and the SENS agenda," in *EMBO Reports* 6, (2005): 1006–1008. Subsequent attempts to debunk de Grey's theories appeared in the February 2005 issue of *MIT Technology Review,* including an editorial labeling him a "troll" that opined, "even if it were possible to "perturb" human biology in the way de Grey wishes, we shouldn't do it." To get a flavor of the man's style, do a YouTube search for "Aubrey de Grey debates" and you will soon appreciate why he drives his critics nuts.

p. 34, *"thirty World Trade Centers* every day*":* Aubrey de Grey, *Ending Aging.* De Grey's ideas and his rather unique worldview are explored quite thoroughly in Jonathan Weiner's excellent *Long For This World: The Strange Science of Immortality* (New York: Ecco Press, 2010), which is well worth a read for those who are interested in the deep cellular biology of aging.

p. 34, *"since roughly 1952":* This was pointed out by the great Leonard Hayflick in an amusing essay. L. Hayflick et al., "Has anyone ever died of old age?" (New York: International Longevity Center–USA, 2003).

p. 35, *"doubles roughly every eight years":* Benjamin Gompertz, "On the Nature of the Function Expressive of the Law of Human Mortality, and on a New Mode of Determining the Value of Life Contingencies," *Philosophical Transactions of the Royal Society,* London, published 1 January 1825.

p. 38, *"median ideal lifespan":* Pew Research Center, "Living to 120 and Beyond: Americans' Views on Aging, Medical Advances, and Radical Life Extension," Washington, D.C., 2013. http://www.pewforum.org/2013/08/06/living-to-120-and-beyond-americans-views-on-aging-medical-advances-and-radical-life-extension/.

Chapter 3: The Fountain of Youthiness

p. 41, *"I am my own experiment"*: You can watch the beginning of Somers's fascinating A4M talk here: www.youtube.com/watch?v=hqst6op9wuI.

p. 43, *"Klatz and Goldman responded by suing"*: The lawsuit was widely covered in the media when it was first filed, by, e.g., the *Chicago Tribune, Inside Higher Ed*, etc.

p. 44, *"Adolf Hitler himself"*: The drug, and others, was administered by Hitler's personal physician, Dr. Theodor Morell, and described in his diaries. His steroid use is described in *Steroids: A New Look at Performance-Enhancing Drugs*, by Rob Beamish (Santa Barbara, CA: Praeger, 2011).

p. 45, *"renegade cancer doctors"*: Both Stanislaw Burzynski and Richard Gonzalez have lengthy and controversial histories that would take a whole chapter to unpack. In a nutshell, Houston-based Burzynski's treatments involve administering so-called "antineoplastons," substances isolated from human urine that he claims can cure many incurable cancers. Because the antineoplastons have not been approved by the FDA (or validated by independent researchers), Burzynski enrolled his patients in clinical trials, allowing them to receive the neoplastons, but FDA inspectors have found numerous problems with his handling of the trials, and have issued multiple warning letters; few if any trial results have ever been published, according to *USA Today*, which has reported extensively on his activities (e.g., "Doctor accused of selling false hope to cancer patients," by Liz Szabo, July 8, 2014). New York–based Gonzalez, on the other hand, treats his patients with a complicated regimen of a customized diet, coffee enemas (for "detoxification") and up to 150 supplements per day. (Sound familiar?) An NIH-sponsored clinical trial of patients with inoperable pancreatic cancer found that patients on Gonzalez's program lived for an average of 4.3 months, versus 14 months for those on traditional chemotherapy; further, the chemo patients reported a better quality of life. He was the subject of a profile by Michael Specter in the *New Yorker*, "The Outlaw Doctor," published February 5, 2001.

p. 45, *"Crazy Talk"*: Weston Kosova and Pat Wingert, "Why Health Advice on 'Oprah' Could Make You Sick," *Newsweek*, May 29, 2009. Well worth a read.

p. 46, *"the massive Women's Health Initiative study"*: For the original paper, see "Risks and Benefits of Estrogen Plus Progestin in Healthy Postmenopausal Women: Principal Results From the Women's Health Initiative Randomized Controlled Trial." *JAMA*.2002;288(3):321-333 (available online free at http://jama.jamanetwork.com/article.aspx?articleid=195120). The study was criticized on many grounds, one of which was that it looked at women who were much older than 50 when they began using hormone replacement, and were thus too old to warrant the increased risk. (The observed benefits of hormone replacement, in the WHI study, included reduced risk of colorectal cancer and hip fractures.) Nevertheless, other major studies of estrogen replacement, in Sweden and in the U.K., found massively increased risk of breast cancer among women taking estrogen and progestin, the usual combination of hormones.

p. 47, *"Another thing that tanked":* P. M. Ravdin et al., "The Decline in Breast Cancer Incidence in 2003 in the United States," *New England Journal Medicine* 2007 April 19;356(16):1670-4. Nancy Krieger of the Harvard School of Public Health wonders whether the heavy promotion of hormone-replacement therapy by the pharmaceutical industry contributed to the steep rise in breast cancer rates during the 1980s: "Hormone therapy and the rise and perhaps fall of US breast cancer incidence rates: critical reflections." (*International Journal of Epidemiology* 2008 June;37(3):627-37). Also, post-WHI, breast cancer rates declined most steeply among the educated, middle- and upper-middle-class white women who were more likely to seek out the treatment (and, after the study, to have then dropped it): N. Krieger et al., "Decline in US breast cancer rates after the Women's Health Initiative: socioeconomic and racial/ethnic differentials." *(American Journal of Public Health,* 2010 April 1;100 Suppl 1:S132-9.)

p. 47, *"Actually, not true":* For well-informed (and skeptical) takes on bioidentical hormone therapy, see A. L. Huntley, "Compounded or confused? Bioidentical hormones and menopausal health." *Menopause International* 2011 March;17(1):16-8.; and Cirigliano M., "Bioidentical hormone therapy: a review of the evidence." *Journal of Women's Health* 2007 June;16(5):600-31. For an excellent layperson's summary of the issues around bioidentical hormone treatment, written by two physicians at the Cleveland Clinic, and complete with charts describing FDA-approved bioidentical options, see Lynn Pattimakiel and Holly Thacker, "Bioidentical Hormone Therapy: Clarifying the Misconceptions," in *Cleveland Clinic Journal of Medicine,* December 2011 pp. 829-836. The pharmacist's indictment is described by journalist Sabrina Tavernise in "First Arrest Made in 2012 Steroid Medication Deaths," *New York Times,* September 4, 2014.

p. 47, *"actual dosages can vary enormously":* e.g., N. A. Yannuzz et al., "Evaluation of compounded bevacizumab prepared for intravitreal injection," *JAMA Ophthalmology.* Published online September 18, 2014.

p. 49, *"a $2 billion business"*: David J. Handelsman, "Global trends in testosterone prescribing, 2000–2011: expanding the spectrum of prescription drug misuse." *Medical Journal of Australia* 2013; 199 (8): 548-551.

p. 49, *"less prone to lying":* True story. Mathias Wibral et al., "Testosterone Administration Reduces Lying in Men." *PLoS One.* 2012; 7(10): e46774. Published online October 10, 2012. The 1941 prostate cancer case was uncovered by Dr. Abraham Morgenthaler, a strong proponent of testosterone therapy and author of books such as *Testosterone For Life* (New York: McGraw-Hill, 2009). The 2010 study that had to be stopped was S. Basaria et al., "Adverse Events Associated with Testosterone Administration," *New England Journal of Medicine* 2010; 363:109-122 July 8, 2010 (available free online at nejm.org). For a good overview of the issues, see this review by researchers at Harvard's Brigham and Women's Hospital, who concluded that "a general policy of testosterone replacement in all older men with age-related decline in testosterone levels is not justified": M. Spitzer et al., "Risks and Benefits of Testosterone Therapy in Older Men," in *Nature Reviews Endocrinology,* 2013 April 16. Information on the NIH clinical trial may be found at trial.org.

p. 52, *"I love fucking":* Ned Zeman, "Hollywood's Vial Bodies," *Vanity Fair,* March 2012. For the AP investigation, see David B. Caruso and Jeff Donn, "Big Pharma Cashes In On HGH Abuse," Associated Press, December 21, 2012. Also worth reading is Brian Alexander's entertaining look at the HGH culture in "A Drug's Promise (Or Not) of Youth," *Los Angeles Times,* July 9, 2006.

p. 53, *"frequented by Rodriguez":* Tim Elfrink, "A Miami Clinic Supplies Drugs to Sports' Biggest Names," *Miami New Times,* January 31, 2013.

p. 55, *"Mintz had once run":* Christopher McDougall, "What if Steroids Were Good For You?" *Best Life,* October 2006. Fascinating profile of Drs. Life and Mintz, published just a few months before Mintz's death.

p. 56, *"a single, small study":* Daniel Rudman et al., "Effects of Human Growth Hormone in Men over 60 Years Old," *New England Journal of Medicine,* 1990; 323:1-6 July 5, 1990. Thirteen years later, the journal revisited the study in an editorial by Mary Lee Vance, "Can Growth Hormone Prevent Aging?" *NEJM* 2003; 348:779-780, which noted that other studies had shown that strength training alone conveyed the same benefits that Rudman had observed from HGH: "Going to the gym is beneficial and certainly cheaper than growth hormone."

p. 58, *"a long list of side effects":* See, e.g., Blackman et al., "Growth hormone and sex steroid administration in healthy aged women and men: a randomized controlled trial." *JAMA* 2002 Nov 13;288(18):2282-92.

p. 59, *"growth-hormone 'knockout' mice":* The strange longevity of animals that lack growth hormone receptors was first observed by Andrzej Bartke in a strain of naturally-occurring mutant mice called the Ames Dwarf; Bartke and others later created a genetically-engineered version of the Ames that is itself the subject of an extensive scientific literature, which is nicely tied together in this review: Andrzej Bartke, "Growth Hormone and Aging: A Challenging Controversy," *Clinical Interventions in Aging,* Dove Press, 2008 December; 3(4): 659–665.

p. 59, *"The bigger dogs produce more IGF-1":* Nathan B. Sutter et al., "A Single *IGF1* Allele Is a Major Determinant of Small Size in Dogs," *Science* 2007 April 6; 316(5821): 112–115.

p. 60, *"It makes me feel good":* "Aging Baby Boomers turn to hormone; some doctors concerned about 'off-label' use of drug," by Sabin Russell, *San Francisco Chronicle,* November 17, 2003. Three months after the story appeared, she was dead: "Cancer took life of noted user of growth hormone," by Sabin Russell, *San Francisco Chronicle,* June 8, 2006.

p. 63, *"in 700 years' time."*: Alex Comfort, *The Biology of Senescence* 3rd ed. (New York: Elsevier, 1964), 1. This is one of my favorite quotes about aging, and its implications have yet to be fully explored. A few other early gerontologists had attempted parabiosis studies, including Clive McCay, who also discovered the life-extending effect of caloric restriction (aka hunger), but Ludwig's was by far the largest and most systematic early study. Frederic Ludwig, "Mortality in Syngeneic Rat Parabionts of Different Age," *Transactions of the New York Academy of Sciences* 1972 Nov;34(7):582-7.

Chapter 4: Yours Sincerely, Wasting Away

p. 67, *"pioneering gerontologist Nathan Shock"*: For background on the BLSA, see Nathan W. Shock et al., "Normal Human Aging: The Baltimore Longitudinal Study of Aging," U.S. Government Printing Office, 1984. It made the papers once in a while, too: Susan Levine, "A New Look at an Old Question: Baltimore research transforms fundamental understanding of aging," *Washington Post,* February 10, 1997; and Nancy Szokan, "Study on Aging Reaches Half-century Mark," *Washington Post,* December 9, 2008.

p. 69, *"most complete medical evaluation that taxpayer money could buy"*: Officially, the BLSA is not intended to replace regular checkups with one's physician, and many of the tests are not considered "diagnostic" in quality, but participants do receive basic results of blood and urine tests, and some others; if study workers detect a potential problem, such as evidence of cancer, they will notify participants.

p. 72, *"The slower you walk"*: There is much research on the issue of "gait speed" in aging, and many studies link walking pace to mortality rates, as well as disability, nursing-home admission, and other very bad things. Here's one: S. Studenski et al., "Gait speed and survival in older adults." *JAMA*. 2011;305:50-58. Luigi Ferrucci, Eleanor Simonsick, and colleagues tie the evidence together in Schrack et al., "The Energetic Pathway to Mobility Loss: An Emerging New Framework for Longitudinal Studies on Aging," in *Journal of the American Geriatric Society*. 2010 October; 58(Suppl 2): S329–S336.

p. 73, *"simple handgrip strength in middle age"*: Taina Rantanen, Jack Guralnick, et al., "Midlife Hand Grip Strength as a Predictor of Old Age Disability." *JAMA*. 1999;281(6):558-560.

p. 77, *"a U-shaped curve"*: The *Economist* summarized this research nicely in a 2010 cover story, "The U-Bend of Life: Why, beyond middle age, people get happier as they get older." December 16, 2010. "Life Begins At 46," proclaimed the cover.

p. 80, *"thousands of Japanese American men in Hawaii"*: Bradley J. Willcox et al., "Midlife Risk Factors and Healthy Survival in Men." *JAMA* 2006;296:2343-2350.

Chapter 5: How to Live to 108 Without Really Trying

p. 85, *"More recently, curcumin":* There is a growing number of studies of curcumin, but the vast majority are in vitro (that is, in the lab dish), or in mice and rats, which do not always equate to human results. What human studies have been done tend to be small, with two dozen subjects or less. For a good overview of human clinical trials to date, see Gupta et al., "Therapeutic roles of curcumin: lessons learned from clinical trials." *AAPS J.* 2013 Jan;15(1):195-218.

p. 85, *"eight* grams *of it each day":* One major issue with curcumin has to do with "bioavailability," how much of the stuff actually gets into our bloodstream; research shows that most of it gets chewed up in the liver, which is one reason my dad takes so much of the stuff. Studies show that you need to take about five grams for it to show up in blood and tissue. Hani et al., "Solubility Enhancement and Delivery Systems of Curcumin an Herbal Medicine: A Review." *Current Drug Delivery,* August 25, 2014. More research into the mechanisms of action of curcumin and its effectiveness in human beings is urgently needed.

p. 87, *"they'll reach for a knish":* Rajpathak et al., "Lifestyle Factors of People With Exceptional Longevity," *Journal of the American Geriatrics Society* 59:8; 1509-12, published online August 2011.

p. 89, *"the men still died sooner":* J. Collerton et al. (2009). "Health and disease in 85 year olds: baseline findings from the Newcastle 85+ cohort study." *British Medical Journal* 339: b4904.

p. 93, *"Studies of Danish twins":* A. M. Herskind et al., "The heritability of human longevity: a population-based study of 2872 Danish twin pairs born 1870–1900," *Human Genetics* 97(3): 319-323 (1996).

p. 94, *"forty-four centenarians":* Barzilai et al., unpublished; personal communication.

p. 94, *"the less CETP you have, the better":* A. E. Sanders, C. Wang, M. Katz, et al., "Association of a Functional Polymorphism in the Cholesteryl Ester Transfer Protein (CETP) Gene With Memory Decline and Incidence of Dementia." *JAMA* 2010;303(2):150-158.

p. 95, *"those drugs have not yet panned out":* Pharma blogger and chemist Derek Lowe thinks CETP inhibitors were a bad bet: http://pipeline.corante.com/archives/2013/01/25/cetp_alzheimers_monty_hall_and_roulette_and_goats.php.

p. 95, *"One other possible longevity gene had to do with IGF-1":* S. Milman et al. (2014). "Low insulin-like growth factor-1 level predicts survival in humans with exceptional longevity." *Aging Cell* 13(4): 769-771.

Chapter 6: The Heart of the Problem

p. 103, *"six hundred thousand Americans":* According to the Centers for Disease Control, 596,577 died of heart disease in 2011, versus 576,691 of cancer. Cancer is gaining fast, some experts believe, simply because people are surviving heart disease and living longer. http://www.cdc.gov/nchs/fastats/leading-causes-of-death.htm.

p. 104, *"serious coronary arteriosclerosis":* W. F. Enos et al. (1953). "Coronary disease among United States soldiers killed in action in Korea; preliminary report." *JAMA* 152(12): 1090-1093.

p. 104, *"a single major risk factor":* D. M. Lloyd-Jones et al. (2006). "Prediction of lifetime risk for cardiovascular disease by risk factor burden at 50 years of age," *Circulation* 113(6): 791-798.

p. 106, *"A major study of 136,000 patients":* A. Sachdeva, C. Cannon et al. "Lipid levels in patients hospitalized with coronary artery disease: An analysis of 136,905 hospitalizations," in "Get With The Guidelines," *American Heart Journal,* January 2009 111-117. Russert's condition described in "From a Prominent Death, Some Painful Truths," by Denise Grady, *New York Times,* June 24, 2008.

p. 106, *"certain outward signs of aging":* M. Christoffersen et al. (2014). "Visible age-related signs and risk of ischemic heart disease in the general population: a prospective cohort study." *Circulation* 129(9): 990-998. For reaction time, see G. Hagger-Johnson et al. (2014). "Reaction time and mortality from the major causes of death: the NHANES-III study." *PLoS One* 9(1): e82959; and of course don't miss E. Banks et al. (2013). "Erectile dysfunction severity as a risk marker for cardiovascular disease hospitalisation and all-cause mortality: a prospective cohort study." *PLoS Medicine* 10(1): e1001372.

p. 108, *"ApoB is a much better predictor":* See G. Walldius et al. (2001). "High apolipoprotein B, low apolipoprotein A-I, and improvement in the prediction of fatal myocardial infarction (AMORIS study): a prospective study." *Lancet* 358(9298): 2026-2033; and also McQueen, M. J., et al. (2008). "Lipids, lipoproteins, and apolipoproteins as risk markers of myocardial infarction in 52 countries (the INTERHEART study): a case-control study." *Lancet* 372(9634): 224-233. These are both very large, persuasive studies showing that APoB is a much better predictor than plain vanilla HDL-LDL cholesterol numbers.

p. 110, *"Red meat has long been known":* This is a bit controversial, but major studies link red meat consumption not only to heart disease but diabetes and cancer. Most notably, see Colin Campbell's *The China Study* (Dallas: BenBella Books, 2005). The TMAO study is small but very interesting and has been followed by other studies of meat and the microbiome. R. A. Koeth et al. (2013). "Intestinal microbiota metabolism of L-carnitine, a nutrient in red meat, promotes atherosclerosis." *Nature Medicine* 19(5): 576-585. The processed vs. unprocessed question is answered pretty clearly in J. Kaluza et al. (2014). "Processed and unprocessed red meat consumption and risk of heart failure: prospective study of men." *Circulation Heart Failure* 7(4): 552-557, which found that unprocessed red meats were not linked to heart failure but processed meats were. (Whew!)

p. 111, *"a brand-new condition":* See R. C. Thompson et al. (2013). "Atherosclerosis across 4000 years of human history: the Horus study of four ancient populations." *Lancet* 381(9873): 1211-1222. The account of Ruffer's work is from A. T. Sandison, "Sir Marc Armand Ruffer (1859-1917), Pioneer of Paleopathology," *Medical History*, April 1967; 11(2): 150–156.

p. 114, *"damage from* intrinsic *aging":* The subject of arterial aging is covered, thoroughly and masterfully, by Ed Lakatta of the National Institute of Aging and colleagues in a series of three papers, beginning with E. G. Lakatta and D. Levy (2003). "Arterial and cardiac aging: major shareholders in cardiovascular disease enterprises: Part I: aging arteries: a 'set up' for vascular disease." *Circulation* 107(1): 139-146.

Chapter 7: Baldness as Metaphor

p. 120, *"Hair loss... affects women, too":* Desmond C. Gan and Rodney D. Sinclair, "Prevalence of Male and Female Pattern Hair Loss in Maryborough," *Journal of Investigative Dermatology Symposium Proceedings* (2005) 10, 184–189. Also see M. P. Birch et al., "Hair density, hair diameter, and the prevalence of female pattern hair loss," *British Journal of Dermatology,* 2001 Feb;144(2): 297-304. I read this stuff so you don't have to.

p. 121: *"key culprit in hair loss":* L. A. Garza et al. (2012). "Prostaglandin D2 inhibits hair growth and is elevated in bald scalp of men with androgenetic alopecia." *Science Translational Medicine* 4(126): 126a–134.

p. 121, *"a bunch of tiny jabs":* M. Ito et al. (2007). "Wnt-dependent de novo hair follicle regeneration in adult mouse skin after wounding." *Nature* 447(7142): 316-320.

p. 122, *"to make room for the next generation":* The evolution of aging is discussed in several articles, and books, but one of the better summaries (particularly of Weissmann's theory) is Michael R. Rose et al. (2008), "Evolution of Ageing since Darwin," *Journal of Genetics*, 87, 363–371; the same subject is treated in D. Fabian and T. Flatt (2011) "The Evolution of Aging," in *Nature Education Knowledge* 3(10):9.

p. 122–123, *"The idea of group-based selection":* Some theorists have revived the idea that our aging might serve some sort of evolutionary purpose—in particular, population control. One fact that has been widely observed is that an abundance of food actually makes most animals *more* likely to die young. This holds true up and down the tree of life, from single-celled organisms all the way up to the people you see in Walmart. Could that be some sort of mechanism to keep us, and things like locusts, from overpopulating the earth and eating everything in sight? Perhaps. Although for 99 percent of human history, this has not been a problem.

p. 123, *"What struck Haldane as odd":* J. B. S. Haldane, "The Relative Importance of Principal and Modifying Genes in Determining Some Human Diseases," in *New Paths in Genetics*. London, G. Allen & Unwin Ltd. 1941.

p. 125, *"the fact that white people get tan":* Zeron-Medina et al. "A Polymorphic p53 Response Element in KIT Ligand Influences Cancer Risk and Has Undergone Natural Selection." *Cell* 155(2): 410-422.

p. 126, *"a gene called daf-2"*: C. Kenyon, J. Chang, E. Gensch, A. Rudner, R. Tabtiang (1993). "A *C. elegans* mutant that lives twice as long as wild type," *Nature* 366 (6454): 461–464. Kenyon later described the process of discovery in C. Kenyon, (2011). "The first long-lived mutants: discovery of the insulin/IGF-1 pathway for ageing." *Philosophical Transactions of the Royal Society B: Biological Sciences* 366(1561): 9-16.

p. 127, *"the long-lived worms were all but extinct"*: Nicole L. Jenkins et al., "Fitness cost of increased lifespan in *C. elegans,*" *Proceedings of the Royal Society London B* (2004) 271, 2523–2526.

p. 127, *"before the advent of birth control"*: V. Tabatabaie et al. (2011). "Exceptional longevity is associated with decreased reproduction." *Aging (Albany NY)* 3(12): 1202-1205.

p. 131, *"time to find a new line of work"*: Austad recounted his dealings with Orville (and the possums) in "Taming Lions, Unleashing a Career," *Science Aging Knowledge Environment*, 27 March 2002, Issue 12, p. vp3.

p. 132, *"the 'disposable soma' theory"*: Kirkwood explains his theory, and much else about aging, in his excellent book *Time of Our Lives: The Science of Human Aging* (New York: Oxford University Press, 1999). The original "disposable soma" paper is at T. B. Kirkwood, *Evolution of ageing.* 1977. *Nature* 170(5635) 201-4. In more recent years the theory has come under attack for various shortcomings, but few scientists disagree with its big-picture conclusion, that lifespan and reproduction exist in a delicate balance.

p. 133, *"The clam, nicknamed Ming"*: Ming's discovery was reported in this very dry paper on oceanography (by Paul Butler and others) titled "Variability of marine climate on the North Icelandic Shelf in a 1357-year proxy archive based on growth increments in the bivalve *Arctica islandica.*" *Palaeogeography, Palaeoclimatology, Palaeoecology* Volume 373, 1 March 2013, pages 141–151; her (his?) death was widely reported, e.g., in "New Record: World's Oldest Animal was 507 Years Old," by Lise Brix, ScienceNordic.com, November 6, 2013.

p. 134, *"what he calls Methuselah's Zoo"*: S. N. Austad, "Methuselah's Zoo: How Nature provides us with clues for extending human healthspan," in *Journal of Comparative Pathology*, 2010 January; 142(Suppl 1): S10–S21.

p. 135, *"bat cells withstood stress"*: A. B. Salmon et al. (2009). "The long lifespan of two bat species is correlated with resistance to protein oxidation and enhanced protein homeostasis." *Journal of the Federation of American Societies of Experimental Biology* 23(7): 2317-2326.

Chapter 8: The Lives of Our Cells

p. 139, *"No one dared question his work"*: J. A. Witkowski, "Alexis Carrel and the Mysticism of Tissue Culture," *Medical History*, 1979, 23: 279-296. More recent historians have been a bit less condemnatory of Carrel, who had some good ideas along with the bad.

p. 141, *"Far from being immortal":* L. Hayflick, "The Limited in vitro Lifetime of Human Diploid Cell Strains," *Experimental Cell Research* 67: 614-36 (1965).

p. 142, *"WI-38 proved to be the most durable and useful cell line":* The story of Hayflick and WI-38, and his role in the science of tissue culture, is told very well by journalist Meredith Wadman in "Medical Research: Cell Division," *Nature* 498, 422–426 (27 June 2013). Hayflick's settlement with the government was described by Philip Boffey, "The Fall and Rise of Leonard Hayflick, Biologist whose Fight With U.S. Seems Over," *New York Times*, January 19, 1982.

p. 142, *"the center of the abortion controversy":* The Vatican's objections were summarized here: http://www.immunize.org/concerns/vaticandocument.htm. Many fundamentalist Christians still refuse certain vaccinations on these grounds.

p. 143, *"his assistants had been replenishing them":* L. Hayflick, interview, March 1, 2013; for further treatment of Carrel's influence on the study of aging, see H. W. Park, (2011), "'Senility and death of tissues are not a necessary phenomenon': Alexis Carrel and the origins of gerontology." *Uisahak* 20(1): 181-208.

p. 145, *"These telomeres, as they were called":* The telomeres-telomerase story is well retold by Carol Greider in "Telomeres and senescence: The history, the experiment, the future," *Current Biology* Vol 8, Issue 5, 26 February 1998, pages R178–R181.

p. 146, *"telomere length and overall mortality":* "A. L. Fitzpatrick, R. A. Kronmal, M. Kimura, J. P. Gardner, B. M. Psaty et al. (2011)Leukocyte telomere length and mortality in the Cardiovascular Health Study." *Journals of Gerontology A: Biological Sciences/Medical Science* 66: 421–429; Also E. S. Epel et al. (2004). "Accelerated telomere shortening in response to life stress." *Proceedings of the National Academy of Sciences* 101(49): 17312-17315.

p. 147, *"if you control for unhealthy behaviors":* M. Weischer et al. (2014). "Telomere shortening unrelated to smoking, body weight, physical activity, and alcohol intake: 4,576 general population individuals with repeat measurements 10 years apart." *PLoS Genetics* 10(3): e1004191.

p. 147, "a widely publicized study": Mariela Jaskelioff et al., "Telomerase reactivation reverses tissue degeneration in aged telomerase deficient mice," *Nature* 2011 January 6; 469(7328): 102–106.

p. 148, *"more liver tumors than the control mice":* Bruno Bernardes de Jesus et al., "The telomerase activator TA-65 elongates short telomeres and increases health span of adult/old mice without increasing cancer incidence," *Aging Cell* (2011) 10, 604–621. Despite the title, the mice treated with TA-65 were 30 percent more likely to develop lymphoma *and* cancer in their livers (p. 615), but because the study involved just 36 animals in total, the finding was deemed not statistically significant by the authors.

p. 150, *"the twenty-five-year-long Rancho Bernardo study":* J. K. Lee et al. (2012). "Association between Serum Interleukin-6 Concentrations and Mortality in Older Adults: The Rancho Bernardo Study." *PLoS One* 7(4): e34218.

p. 152, *"nestled against a Marin County hillside":* Heading north from San Francisco on U.S. 101, look to your left just past the exit for Novato, and you'll see the Buck Institute up on the hill. Easily the most spectacular research institute I've ever seen, but in 2013 it was nearly bankrupted by lawsuits from creditors of the defunct Lehman Brothers investment firm. (Long story: "Lehman Reaches from beyond Grave Seeking Millions from Nonprofits," by Martin Z. Braun, Bloomberg.com, May 24, 2013.)

p. 152, *"the senescence-associated secretory phenotype":* This is one of the most important concepts in cellular aging, and it has been found to have broad implications for physiology and health. Campisi's original SASP paper is here: J. P. Coppe et al. (2008). "Senescence-associated secretory phenotypes reveal cell-nonautonomous functions of oncogenic RAS and the p53 tumor suppressor." *PLoS Biology* 6(12): 2853-2868. For less thorny reading, try J. Campisi et al. (2011), "Cellular senescence: a link between cancer and age-related degenerative disease?" *Seminars in Cancer Biology* 21(6): 354-359. Or better yet, search on YouTube for "Senescent cells Campisi"—she's a very good speaker.

p. 153, *"patients who had been treated for HIV":* The observed rapid aging of the drug-treated HIV population raises its own issues, but has also shed insight on the nature of the aging process itself, in particular the relationship of the immune system to overall aging. J. B. Kirk and M. B. Goetz (2009), "Human immunodeficiency virus in an aging population, a complication of success," *Journal of the American Geriatrics Society* 57(11): 2129-2138.

p. 154, *"senescent-cell-zapping drug":* The original paper is available free on the *Nature* website: D. J. Baker et al. (2011), "Clearance of p16Ink4a-positive senescent cells delays ageing-associated disorders," *Nature* 479(7372): 232-236. For a simpler account, try "Cell-Aging Hack Opens Longevity Research Frontier," by Brandon Keim, Wired.com, November 2, 2011.

Chapter 9: Phil vs. Fat

p. 157–158, *"Waist circumference also expands":* Waist circumference, also known as your waist size, is one of the most important "biomarkers" there is, far more important than BMI; numerous studies have linked waist circumference of more than 1 meter (in men) to all kinds of poor health outcomes (for women, the threshold is more like 36 inches). As a rule, your waist should be less than one-half your height: M. Ashwell et al. (2014), "Waist-to-height ratio is more predictive of years of life lost than body mass index," *PLoS One* 9(9): e103483.

p. 158, *"very high end of 'normal'":* Body-fat percentage ranges are from the American Council on Exercise: http://www.acefitness.org/acefit/healthy-living-article/60/112/what-are-the-guidelines-for-percentage-of/. Note that these are only averages; some sports scientists advocate still lower levels for both men and women, depending on what sport they pursue. http://www.humankinetics.com/excerpts/excerpts/normal-ranges-of-body-weight-and-body-fat.

p. 161, *"14 percent of cancer deaths in men":* E. E. Calle, C. Rodriguez, K. Walker-Thurmond, M. J. Thun. "Overweight, obesity, and mortality from cancer in a prospectively studied cohort of U.S. adults." *New England Journal of Medicine* 2003 Apr 24;348(17):1625-38. Subsequent research has muddied the issue a bit, with some studies concluding that the ideal weight, mortality-wise, is somewhere around BMI of 25, on the edge of overweight; but while it may be better to be slightly overweight than underweight, more studies show that outright obesity is always associated with higher risk of disease and death.

p. 162, *"taking in just ten more calories":* The *Lancet* report also explained why it then becomes so hard to lose weight, because a successful diet requires cutting back by 250 calories per day, or more. (Good-bye, Hershey bar in the afternoon.) Kevin D. Hall et al., "Quantification of the effect of energy imbalance on bodyweight," *Lancet*, Volume 378, Issue 9793, pp. 826–837, 27 August 2011.

p. 163, *"The fat is the problem.":* One huge study found that even for people of normal weight, excess visceral fat increased their risk of death. More cheery reading for you: T. Pischon et al. (2008), "General and Abdominal Adiposity and Risk of Death in Europe," *New England Journal of Medicine* 359(20): 2105-2120.

p. 165, *"Not all fat is just fat":* This is one of my favorite studies ever: Researchers cut out the animals' visceral fat, and they lived lots longer. R. Muzumdar et al. (2008), "Visceral adipose tissue modulates mammalian longevity," *Aging Cell* 7(3): 438-440.

p. 165, *"only about half of diabetic patients":* In fairness, more patients are being told to exercise by their doctors; percentages increased from 2000 to 2010 across all categories and age groups. But half is still not everyone, and exercise has been shown to be the most powerful intervention against diabetes. Patricia Barnes, National Center for Health Statistics Data Brief no. 86, February 2012. http://www.cdc.gov/nchs/data/databriefs/db86.pdf.

p. 168–169, *"A. B. was a twenty-seven-year-old Scotsman":* W. K. Stewart and Laura W. Fleming, "Features of a Successful Therapeutic Fast of 382 Days' Duration," *Postgraduate Medical Journal*, March 1973; 49(569): 203–209. Maybe my second-favorite study. More recently, the evolutionary biologist John Speakman used a model of total starvation to call into question the "thrifty gene hypothesis," which says that all humans are predisposed to obesity; why, he asked, are we therefore not all fat? J. R. Speakman and K. R. Westerterp (2013), "A mathematical model of weight loss under total starvation: evidence against the thrifty-gene hypothesis," *Disease Models & Mechanisms* 6(1): 236-251.

Chapter 10: Pole Vaulting into Eternity

p. 179, *"to find out just what it is"*: John Jerome, *Staying With It* (New York: Viking, 1984), 219. Jerome died in 2002, at age 70, of lung cancer.

p. 180, *"They were miraculously rejuvenated."*: Langer's experiment was never published, except as a chapter in an obscure book; the results were too far out of the mainstream for 1981. In 2010, it was the basis of a BBC special, featuring aging celebrities. Langer was the subject of a recent *New York Times Magazine* profile, "The Thought That Counts," by Bruce Grierson, October 26, 2014.

p. 181, *"the equivalent of a brisk walk"*: S. C. Moore et al. (2012). "Leisure time physical activity of moderate to vigorous intensity and mortality: a large pooled cohort analysis." *PLoS Medicine* 9(11): e1001335. There is a raging debate over how much exercise constitutes "too much," fueled by numerous studies by James O'Keefe, a cardiologist in Kansas City, who argues that long-term endurance exercise delivers proportionally fewer benefits, and at the cost of possible damaging changes to the heart (e.g., J. H. O'Keefe et al. (2012). "Potential adverse cardiovascular effects from excessive endurance exercise." *Mayo Clinic Proceedings* 87(6): 587-595. For most Americans, however, the problem is not too much exercise, but too little.

p. 182, *"just as effective as the medications"*: Huseyin Naci and John Ioannidis, "Comparative effectiveness of exercise and drug interventions on mortality outcomes: metaepidemiological study." *BMJ* 2013;347:f5577 (published 1 October 2013).

p. 183, *"National records also bear this out"*: See www.mastersrankings.com.

p. 184, *"Even retiring from working"*: Dhaval Dave, Inas Rashad, and Jasmina Spasojevic, "The Effects of Retirement on Physical and Mental Health Outcomes," NBER Working Paper No. 12123. March 2006, January 2008. JEL No. I1,J0. http://www.nber.org/papers/w12123.

p. 186, *"because of something that circulates in old blood"*: Irina M. Conboy et al., "Rejuvenation of aged progenitor cells by exposure to a young systemic environment." *Nature* 433, 760-764 (17 February 2005). There will be lots more on the fascinating science of parabiosis later in the book.

p. 189, *"a male named Charlie"*: Charlie won something called the "Reversal Prize," awarded by the Methuselah Foundation for the longest extension of lifespan of a mouse. His handler, Sandy Keith, was awarded the prize in 2004. NIA scientist Mark Mattson has questioned whether standard captivity conditions, where mice have unlimited access to food but no chance to exercise or socialize, have skewed study results by making the mice unhealthy: B. Martin, S. Ji, S. Maudsley, and M. P. Mattson, "'Control' Laboratory Rodents Are Metabolically Morbid: Why It Matters," *Proceedings of the National Academy of Sciences USA* 107, no. 14 (April 6 2010): 6127-33.

p. 190, *"more than 5.3 million premature deaths each year"*: I. M. Lee et al. (2012), "Effect of physical inactivity on major non-communicable diseases worldwide:

an analysis of burden of disease and life expectancy," *Lancet* 380(9838): 219-229. Other scientists have picked up the ball, describing inactivity itself as a disease or, better, a dangerous activity on par with smoking. B. K. Pedersen (2009). "The diseasome of physical inactivity—and the role of myokines in muscle–fat cross talk," *The Journal of Physiology* 587(23): 5559-5568.

p. 191, *"The primary signaling factor they identified":* B. K. Pedersen and M. A. Febbraio, "Muscles, exercise and obesity: skeletal muscle as a secretory organ," *Nature Reviews Endocrinology*, advance online publication April 3, 2012.

p. 194, *"analyzed the 'gene expression' patterns:* Simon Melov, et al., "Resistance Exercise Reverses Aging in Human Skeletal Muscle," *PLoS ONE* 2(5): e465. Tarnopolsky's subsequent paper, looking at mitochondrial DNA mutation in mice, and its reversal through exercise, is at A. Safdar et al. (2011), "Endurance exercise rescues progeroid aging and induces systemic mitochondrial rejuvenation in mtDNA mutator mice," *Proceedings of the National Academy of Sciences* 108(10): 4135-4140.

p. 199, *"Within eight generations, they found distinct differences":* M. D. Roberts et al. (2014), "Nucleus accumbens neuronal maturation differences in young rats bred for low versus high voluntary running behaviour," *Journal of Physiology* 592(Pt 10): 2119-2135.

p. 201, *"A little bit of walking kept many of them out of the nursing home":* M. Pahor, J. M. Guralnik, W. T. Ambrosius et al., "Effect of Structured Physical Activity on Prevention of Major Mobility Disability in Older Adults: The LIFE Study Randomized Clinical Trial," *JAMA*. 2014;311(23):2387-2396.

Chapter 11: Starving for Immortality

p. 205, *"He titled it* Discorsi della vita sobria*":* My edition, purchased online, was titled simply *How to Live Long* (New York: Health Culture, 1916).

p. 206, *"McCay's resulting paper":* C. M. McCay, and Mary Crowell, (1935), "The effect of retarded growth upon the length of life span and upon the ultimate body size," *Nutrition* 5(3): 155-171. A classic.

p. 207, *"McCay's own special recipe for rat chow":* For more details on Clive McCay's very interesting life, see his wife Jeanette's autobiography, *Clive McCay, Nutrition Pioneer: Biographical Memoirs by His Wife* (Charlotte Harbor, FL: Tabby House, 1994). Also helpful was a doctoral dissertation by historian Hyung Wook Park (2010), "Longevity, aging, and caloric restriction: Clive Maine McCay and the construction of a multidisciplinary research program." *Historical Studies in the Natural Sciences* 40(1): 79-124.

p. 208, *"you could accomplish all kinds of things in this world":* The trailer for Rowland's documentary on Walford, *Signposts of Dr. Roy Walford*, is at www.youtube.com/watch?v=K-PzhyTlODc.

p. 209, *"caloric restriction might actually be slowing the aging process itself"*: Their seminal paper is R. Weindruch, and R. L. Walford (1982), "Dietary restriction in mice beginning at 1 year of age: effect on life-span and spontaneous cancer incidence," *Science* 215(4538): 1415-1418. They later wrote a whole book about caloric restriction that you probably do not want to read.

p. 210, *"which still strikes Finch as 'statistically unlikely.'"*: Walford's good friend Caleb Finch penned a biographical remembrance, "Dining With Roy," *Experimental Gerontology* 39 (2004) 893–894), from which some of these details are drawn. Still more come from recollections kindly shared by both Finch and Rick Weindruch.

p. 211, *"Funded by the eccentric oil heir Ed Bass"*: Good retrospective feature on Biosphere 2 by Tiffany O'Callaghan, "Biosphere 2: Saving the world within a world," *New Scientist*, July 31, 2103.

p. 212, *"The supply of bananas, the tastiest item on the menu"*: Many details about life inside Biosphere come from Jane Poynter, *The Human Experiment: Two Years and Twenty Minutes Inside Biosphere 2* (New York: Basic Books, 2009).

p. 212, *"they had the best blood he had ever seen"*: R. L. Walford et al. (1992) "The calorically restricted low-fat nutrient-dense diet in Biosphere 2 significantly lowers blood glucose, total leukocyte count, cholesterol, and blood pressure in humans," *Proceedings of the National Academy of Sciences* 89(23): 11533-11537. The Biosphere came closest to fulfilling the suggestion, of a former director of the National Institute on Aging, that caloric restriction be studied (forcibly) on prisoners.

p. 214, *"Walford fell into a deep depression"*: The gruesome aftermath of the Biosphere, for Walford at least, is recounted rather clinically in a paper by some of his colleagues: B. K. Lassinger, C. Kwak, R. L. Walford, and J. Jankovic (2004), "Atypical parkinsonism and motor neuron syndrome in a Biosphere 2 participant: A possible complication of chronic hypoxia and carbon monoxide toxicity?" *Movement Disorders*, 19: 465–469.

p. 215, *"he touted the benefits of caloric restriction to Alan Alda"*: The man did not look well: www.youtube.com/watch?v=9jvqNG1g62Y.

p. 216, *"These genes, dubbed sirtuins"*: The sirtuin-discovery story is well recounted by Guarente himself in *Ageless Quest: One Scientist's Search for Genes That Prolong Youth.* (Cold Spring Harbor, NY: Cold Spring Harbor Laboratory Press, 2003).

p. 217, *"they were fitter, faster, and lots better-looking"*: J. A. Baur et al. (2006), "Resveratrol improves health and survival of mice on a high-calorie diet," *Nature* 444(7117): 337-342. The study made the *Times* front page with the headline, "Yes, Red Wine Holds Answer. Check Dosage," by Nicholas Wade, November 2, 2006.

p. 218, *"those who consume red wine"*: The literature on red wine is fascinating, even inspiring; multiple large European studies have shown huge health benefits associated with red wine consumption, and not just the one or two glasses per day that American doctors recommend, but more like three. J. P. Broustet, "Red Wine and Health," *Heart* 1999;81:459-460.

p. 219, *"He hadn't given up"*: In 2014 Sinclair reported results of a *new* chemical sirtuin activator, which seemed to work even better than resveratrol. The unfortunate part is that the chemical, nicotine mononucleotide, currently costs about a thousand bucks per gram. Brace yourself: P. Ana Gomes et al. (2013), "Declining NAD+ Induces a Pseudohypoxic State Disrupting Nuclear-Mitochondrial Communication during Aging," *Cell* 155(7): 1624-1638.

p. 221, *"decades younger than their chronological age"*: Fontana's results with the "Cronies" (Caloric Restriction with Optimal Nutrition), including Dowden, have been reported and analyzed in a long series of studies, beginning with L. Fontana et al., "Long-term calorie restriction is highly effective in reducing the risk for atherosclerosis in humans." *Proceedings of the National Academy of Sciences* 2004 April 7; 101(17):6659-63.

p. 222, *"The hungry monkeys were far healthier"*: R. J. Colman et al. (2009), "Caloric restriction delays disease onset and mortality in rhesus monkeys," *Science* 325(5937): 201-204.

p. 223, *"the 'dieting' monkeys had* not *lived longer"*: J. A. Mattison et al. (2012), "Impact of caloric restriction on health and survival in rhesus monkeys from the NIA study," *Nature* 489(7415): 318-321. For a more readable commentary, published at the same time, see Steven Austad, "Mixed Results for Dieting Monkeys," *Nature* (same issue).

p. 224, *"close to an institutional disaster"*: Although it had become almost dogma that CR would extend lifespan all the time, there had been anomalous CR results in the past. Steven Austad had tried it in actual wild mice, captured in a barn in Idaho, rather than the usual, genetically-standardized laboratory mice. It had minimal effect on their lifespans. In another, more structured study, Jim Nelson of the University of Texas tried it in 40 different cross-bred mouse strains, and found that in a third of the breeds, dietary restriction actually shortened their lives. So clearly, it doesn't work for everyone all the time. Even monkeys.

p. 224, *"better to be a little bit overweight"*: The relationship of BMI to longevity (or mortality) is the subject of yet another thorny, contentious debate. It's called the "obesity paradox," the observation that being slightly overweight and even mildly obese is actually associated with living slightly longer—not only for the population as a whole, but for people who actually have hypertension and diabetes. It would take pages to unpack this, but the latest salvo appears to discredit the "paradox," showing that being obese and diabetic (which the Wisconsin monkeys certainly were) is not good: D. K. Tobias et al. (2014), "Body-Mass Index and Mortality among Adults with Incident Type 2 Diabetes," *New England Journal of Medicine* 370(3): 233-244.

p. 226, *"it's not surprising they had different outcomes"*: Not to be defeated, the Wisconsin team came back with a paper exploring, in detail, the many subtle differences that may have led to the drastically different outcomes of the two studies, not only diet but also the genetic background of the monkeys, their age when the respective studies were started, etc. R. J. Colman et al. (2014), "Caloric restriction reduces age-related and all-cause mortality in rhesus monkeys," *Nature Communications* 5: 3557. If that's still not enough for you, dive into Michael Rae's exhaustive exegesis, "CR in Nonhuman Primates: A Muddle for Monkeys, Men, and Mimetics," posted at www.sens.org on May 6, 2013. If nothing else, you will appreciate the extent to which the Whole Foods (NIH) Monkeys turned the field on its ear.

Chapter 12: What Doesn't Kill You

p. 230, *"His essay about cold-water showering":* http://gettingstronger.org/2010/03/cold-showers/. Remember, the first minute is the worst.

p. 233, *"cold water might help increase their longevity":* For a readable overview on nematode worms and cold, see B. Conti and M. Hansen (2013), "A cool way to live long," *Cell* 152(4): 671-672.

p. 234, *"After months of regular cold swimming":* A. Lubkowska et al. (2013), "Winter-swimming as a building-up body resistance factor inducing adaptive changes in the oxidant/antioxidant status," *Scandinavian Journal of Clinical and Laboratory Investigations,* March 20, 2013 [epub ahead of print]. Think about this before you chicken out of that cold shower. Katharine Hepburn's year-round swimming was documented by her biographer Charles Higham, in *Kate: The Life of Katharine Hepburn* (New York: W. W. Norton, 2004 [First published 1975]). HTFU.

p. 234, *"exposure to cold water could activate brown fat":* Paul Lee et al. "Irisin and FGF21 Are Cold-Induced Endocrine Activators of Brown Fat Function in Humans," *Cell Metabolism,* Volume 19, Issue 2, 4 February 2014, 302-309.

p. 237, *"small doses of heat are also beneficial":* There is an extensive literature on heat-shock proteins, but one of the first to connect the heat-shock response, and indeed the concept of hormesis itself, to longevity was Suresh I. Rattan, in a review that garnered an extremely hostile response but is now widely accepted: S. I. Rattan, "Applying hormesis in aging research and therapy," *Human & Experimental Toxicology.* 2001 Jun;20(6):281-5; discussion 293-4.

p. 239, *"people who were lonely":* S. W. Cole et al. (2007), "Social regulation of gene expression in human leukocytes," *Genome Biology* 8(9): R189.

p. 240, *"Free radicals flashed through my mind":* "An Interview with Dr. Denham Harman," *Life Extension,* February 1998.

p. 242, *"Antioxidants didn't really seem to extend lifespan":* For an excellent breakdown of this subject, see "The Myth of Antioxidants," by Melinda Wenner Moyer, *Scientific American* 308, 62 - 67 (2013); published online January 14, 2013.

p. 242, *"antioxidant supplements have had a mixed track record":* G. Bjelakovic et al. (2007), "Mortality in randomized trials of antioxidant supplements for primary and secondary prevention: Systematic review and meta-analysis," *JAMA* 297(8): 842-857.

p. 243, *"worse than useless":* The supplements-in-exercise study is at M. Ristow, "Antioxidants prevent health-promoting effects of physical exercise in humans." *Proceedings of the National Academy of Sciences of the USA* 106: 8665–8670, 2009. Ristow and others review that and similar studies (with mixed results) in Mari Carmen Gomez-Cabrera et al., "Antioxidant supplements in exercise: worse than useless?" *American Journal of Physiology—Endocrinology and Metabolism* 15 February 2012 Vol. 302no. E476-E477.

p. 245, *"essential signaling molecules":* M. Ristow, and S. Schmeisser (2011), "Extending life span by increasing oxidative stress," *Free Radical Biology in Medicine* 51(2): 327-336. Also see M. Ristow (2014), "Unraveling the truth about antioxidants: mitohormesis explains ROS-induced health benefits," *Nature Medicine* 20(7): 709-711.

p. 246, *"they rarely if ever see the sun":* Don't miss the fantastic essay about the naked mole rat by Eliot Weinberger in *Karmic Traces, 1993-1999* (New York: New Directions Publishing, 2000).

p. 248, *"they are better designed to handle stress":* K. N. Lewis et al. (2012). "Stress resistance in the naked mole-rat: the bare essentials—a mini-review." *Gerontology* 58(5): 453-462. A good place to start, and my favorite bad-pun journal article title ever. Also see Y. H. Edrey et al. (2011), "Successful aging and sustained good health in the naked mole rat: a long-lived mammalian model for biogerontology and biomedical research," *ILAR J* 52(1): 41-53.

p. 249, *"a kind of cave-dwelling salamander":* The evolutionary biologist John Speakman uses the olm to help further dismantle the oxidative-stress theory of aging, in "The free-radical damage theory: Accumulating evidence against a simple link of oxidative stress to ageing and lifespan," *Bioessays* 33: 255–259 (2011).

p. 250, *"sequence the naked mole rat genome":* E. B. Kim et al. (2011), "Genome sequencing reveals insights into physiology and longevity of the naked mole rat," *Nature* 479(7372): 223-227.

Chapter 13: Fast Forward

p. 254, *"simply feeding his lab animals every other day":* Anton J. Carlson and Frederic Hoelzel, "Apparent Prolongation of the Lifespan of Rats by Intermittent Fasting," *Journal of Nutrition*, March 1946 31:363-75; the Spanish nursing-home study is described by Johnson et al., (2006), "The effect on health of alternate day calorie restriction: eating less and more than needed on alternate days prolongs life." *Medical Hypotheses* 67(2): 209-211. Longo's take: "The question is, did the 60 that were fed every other day wish they were dead? Probably yes."

p. 255, *"their asthma symptoms also cleared up"*: J. B. Johnson et al. (2007), "Alternate day calorie restriction improves clinical findings and reduces markers of oxidative stress and inflammation in overweight adults with moderate asthma," *Free Radical Biology in Medicine* 42(5): 665-674.

p. 255, *"Studies of Muslims during Ramadan"*: For a good review of studies on religious fasting and health, check out John F. Trepanowski et al., "Impact of caloric and dietary restriction regimens on markers of health and longevity in humans and animals: a summary of available findings," *Nutrition Journal* 2011, 10:107 (http://www.nutritionj.com/content/10/1/107).

p. 256, *"brain-derived neurotrophic factor"*: Mark Mattson has amassed a string of fascinating studies on diet and fasting, but the brain work is the most interesting. In this study, he and his team found that intermittent fasting improved glucose handling AND protected neurons, independent of the total number of calories consumed: R. M. Anson et al. (2003), "Intermittent fasting dissociates beneficial effects of dietary restriction on glucose metabolism and neuronal resistance to injury from calorie intake," *Proceedings of the National Academy of Sciences* 100(10): 6216-6220.

p. 257, *"there's no one 'right way' to do intermittent fasting"*: The 8-hour feeding "window" study is at M. Hatori et al. (2012), "Time-restricted feeding without reducing caloric intake prevents metabolic diseases in mice fed a high-fat diet," *Cell Metabolism* 15(6): 848-860. The last couple of years has seen a flood of fasting-based diet books hit the market, from *The 8-Hour Diet* (loosely based on Panda's work) to *The Every-Other-Day Diet: The Diet That Lets You Eat All You Want (Half the Time) and Keep the Weight Off*, by University of Illinois at Chicago professor Krista Varady (New York: Hyperion, 2013). Then there's the UK bestseller *The Fast Diet*, by Michael Mosley (New York: Atria Books, 2013), which advocates fasting two days out of seven (and which Longo says has no basis in any research whatsoever). Fasting, in other words, is trendy.

p. 259, *"blocking the TOR pathway caused his yeast to live three times longer"*: The Longo group's breakthrough yeast paper is Paola Fabrizio et al., "Regulation of Longevity and Stress Resistance by Sch9 in Yeast," *Science* 13 April 2001: Vol. 292 no. 5515 pp. 288-290. Way more interesting than this sounds.

p. 261–262, *"In some patients, the chemo also appeared to be more effective"*: Both the mouse and human fasting plus chemotherapy trials are described in F. M. Safdie et al. (2009), "Fasting and cancer treatment in humans: A case series report," *Aging* (Albany NY) 1(12): 988-1007. Larger clinical trials are ongoing, with results expected in later 2015.

p. 263, *"he took his precious soil fungus with him"*: The discovery and development of rapamycin, by Indian-born Canadian scientist Suren Sehgal, is one of the great serendipitous tales in modern biology. The story is well told in "Rapamycin's Resurrection: A New Way to Target the Cell Cycle," in *Journal of the National Cancer Institute*, October 17, 2001.

p. 263–64, *"rapamycin had significantly extended the lifespan of mice":* D. E. Harrison et al. (2009), "Rapamycin fed late in life extends lifespan in genetically heterogeneous mice," *Nature* 460(7253): 392-395. The *Times* buried the story, and inaccurately described rapamycin as an "antibiotic."

p. 265, *"rapamycin actually reversed cardiac aging in elderly mice":* J. M. Flynn et al. (2013), "Late-life rapamycin treatment reverses age-related heart dysfunction," *Aging Cell* 12(5): 851-862.

p. 266, *"rapamycin would likely extend lifespan in mammals":* M. V. Blagosklonny, (2006), "Aging and immortality: quasi-programmed senescence and its pharmacologic inhibition," *Cell Cycle* 5(18): 2087-2102.

p. 267, *"it's more like a program gone wrong":* The term "hyperfunction" was first used by London gerontologist David Gems; see D. Gems, and Y. de la Guardia (2013), "Alternative Perspectives on Aging in Caenorhabditis elegans: Reactive Oxygen Species or Hyperfunction?" *Antioxidant Redox Signalling* 19(3): 321-329.

p. 270, *"Called Laron little people":* J. Guevara-Aguirre et al. (2011), "Growth Hormone Receptor Deficiency Is Associated with a Major Reduction in Pro-Aging Signaling, Cancer, and Diabetes in Humans," *Science Translational Medicine* 3(70): 70ra13.

p. 272, *"middle-aged people who had eaten a diet high in dairy and meat":* M. E. Levine et al. (2014), "Low protein intake is associated with a major reduction in IGF-1, cancer, and overall mortality in the 65 and younger but not older population," *Cell Metabolism* 19(3): 407-417.

Chapter 14: Who Moved My Keys?

p. 278, *"significant cognitive decline is already evident":* Archana, Singh-Manoux, Mika Kivimaki, M. Maria Glymour, Alexis Elbaz, Claudine Berr, Klaus P. Ebmeier, Jane E. Ferrie, and Aline Dugravot, "Timing of Onset of Cognitive Decline: Results from Whitehall II Prospective Cohort Study." *BMJ* Vol. 344, 2012. Journal Article. doi:10.1136/bmj.d7622. For the depressing fruit-fly information, thank Hsueh-Cheng Chiang, Lei Wang, Zuolei Xie, Alice Yau, and Yi Zhong. "Pi3 Kinase Signaling Is Involved in Aβ-Induced Memory Loss in Drosophila." *Proceedings of the National Academy of Sciences* 107, no. 15 (April 13, 2010 2010): 7060-65.

p. 278, *"Her name was Auguste D.,":* The whole fascinating story is retold here, complete with original drawings and photographs. Interestingly, Auguste D. is now thought to have suffered from arteriosclerosis of the brain, not Alzheimer's. M. B. Graeber, S. Kosel, R. Egensperger, R. B. Banati, U. Muller, K. Bise, P. Hoff, et al., "Rediscovery of the Case Described by Alois Alzheimer in 1911: Historical, Histological and Molecular Genetic Analysis," *Neurogenetics* 1, no. 1 (May 1997): 73-80.

p. 280, *"In clinical trials in actual patients, they failed to work"*: J. L. Cummings, T. Morstorf, and K. Zhong, "Alzheimer's Disease Drug-Development Pipeline: Few Candidates, Frequent Failures." [In English]. *Alzheimer's Research and Therapeutics* 6, no. 4 (2014): 37. Also "Alzheimer's Theory That's Been Drug Graveyard Facing Test," by Michelle Fay Cortez and Drew Armstrong, *Bloomberg News,* December 12, 2013.

p. 281, *"the better writers' brains were also less gunked-up"*: D. Iacono, W. R. Markesbery, M. Gross, O. Pletnikova, G. Rudow, P. Zandi, and J. C. Troncoso, "The Nun Study: Clinically Silent Ad, Neuronal Hypertrophy, and Linguistic Skills in Early Life." *Neurology* 73, no. 9 (Sep 1 2009): 665-73. The Nun Study generated many fascinating publications (available at https://www.healthstudies.umn.edu/nunstudy/publications.jsp), and also a book: David Snowdon, *Aging with Grace: What the Nun Study Teaches Us About Leading Longer, Healthier, and More Meaningful Lives* (New York: Bantam Books, 2001).

p. 282, *"their brains had all the hallmarks of clinical Alzheimer's"*: I. Driscoll, S. M. Resnick, J. C. Troncoso, Y. An, R. O'Brien, and A. B. Zonderman, "Impact of Alzheimer's Pathology on Cognitive Trajectories in Nondemented Elderly," *Annals of Neurology* 60, no. 6 (Dec 2006): 688-95.

p. 282, *"what's called cognitive reserve"*: Good summary of the topic in Yaakov Stern, "Cognitive Reserve," *Neuropsychologia* 47, no. 10 (August 2009): 2015-28.

p. 283, *"Jet lag ages you."*: A. J. Davidson, M. T. Sellix, J. Daniel, S. Yamazaki, M. Menaker, and G. D. Block, "Chronic Jet-Lag Increases Mortality in Aged Mice," *Current Biology* 16, no. 21 (November 7, 2006): R914-6.

p. 283, *"half of all Alzheimer's cases could actually be prevented"*: D. E. Barnes, and K. Yaffe, "The Projected Effect of Risk Factor Reduction on Alzheimer's Disease Prevalence," *Lancet Neurology* 10, no. 9 (September 2011): 819-28.

p. 284, *"merely walking twenty minutes a day"*: J. Winchester, M. B. Dick, D. Gillen, B. Reed, B. Miller, J. Tinklenberg, D. Mungas, et al., "Walking Stabilizes Cognitive Functioning in Alzheimer's Disease across One Year," *Archives of Gerontololgy and Geriatrics* 56, no. 1 (January–February 2013): 96-103.

p. 285, *"Old blood hurts your brain"*: S. A. Villeda, J. Luo, K. I. Mosher, B. Zou, M. Britschgi, G. Bieri, T. M. Stan, et al., "The Ageing Systemic Milieu Negatively Regulates Neurogenesis and Cognitive Function," *Nature* 477, no. 7362 (September 1, 2011): 90-4.

p. 286, *"Young blood had restored their old brains."*: S. A. Villeda , K. E. Plambeck, J. Middeldorp, J. M. Castellano, K. I. Mosher, J. Luo, L. K. Smith, et al., "Young Blood Reverses Age-Related Impairments in Cognitive Function and Synaptic Plasticity in Mice," *Nature Medicine* 20, no. 6 (June 2014): 659-63.

p. 287, *"to repair their injured muscles."*: I. M. Conboy, M. J. Conboy, A. J. Wagers, E. R. Girma, I. L. Weissman, and T. A. Rando, "Rejuvenation of Aged Progenitor Cells by Exposure to a Young Systemic Environment." *Nature* 433, no. 7027 (February 17, 2005): 760-4.

p. 289, *"It had set back the clock"*: F. S. Loffredo, M. L. Steinhauser, S. M. Jay, J. Gannon, J. R. Pancoast, P. Yalamanchi, M. Sinha, et al., "Growth Differentiation Factor 11 Is a Circulating Factor That Reverses Age-Related Cardiac Hypertrophy," *Cell* 153, no. 4 (May 9 2013): 828-39. The story is also told in "Young Blood," *Science*, September 12, 2014: Vol. 345 no. 6202 pp. 1234-1237.

p. 289, *"The elderly mice had even regained their sense of smell"*: L. Katsimpardi, N. K. Litterman, P. A. Schein, C. M. Miller, F. S. Loffredo, G. R. Wojtkiewicz, J. W. Chen, et al., "Vascular and Neurogenic Rejuvenation of the Aging Mouse Brain by Young Systemic Factors." *Science* 344, no. 6184 (May 9, 2014): 630-4. The muscle-rejuvenation paper is M. Sinha, Y. C. Jang, J. Oh, D. Khong, E. Y. Wu, R. Manohar, C. Miller, et al., "Restoring Systemic Gdf11 Levels Reverses Age-Related Dysfunction in Mouse Skeletal Muscle," *Science* 344, no. 6184 (May 9, 2014): 649-52.

p. 290, *"Old muscles seemed to be rejuvenated by dosing with oxytocin"*: C. Elabd, W. Cousin, P. Upadhyayula, R. Y. Chen, M. S. Chooljian, J. Li, S. Kung, K. P. Jiang, and I. M. Conboy, "Oxytocin Is an Age-Specific Circulating Hormone That Is Necessary for Muscle Maintenance and Regeneration," *Nature Communications* 5 (2014): 4082.

Epilogue: The Death of Death

p. 292, *"It's called cytomegalovirus"*: P. Sansoni, R. Vescovini, F. F. Fagnoni, A. Akbar, R. Arens, Y. L. Chiu, L. Cicin-Sain, et al., "New Advances in Cmv and Immunosenescence," *Experimental Gerontology* 55 (July 2014): 54-62. L. Cicin-Sain, J. D. Brien, J. L. Uhrlaub, A. Drabig, T. F. Marandu, and J. Nikolich-Zugich, "Cytomegalovirus Infection Impairs Immune Responses and Accentuates T-Cell Pool Changes Observed in Mice with Aging," *PLoS Pathogens* 8, no. 8 (2012): e1002849.

p. 293, *"CMV is particularly strongly associated with preclinical cardiovascular disease"*: N. C. Olson, M. F. Doyle, N. S. Jenny, S. A. Huber, B. M. Psaty, R. A. Kronmal, and R. P. Tracy, "Decreased Naive and Increased Memory Cd4(+) T Cells Are Associated with Subclinical Atherosclerosis: The Multi-Ethnic Study of Atherosclerosis." [In English]. *PLoS One* 8, no. 8 (2013): e71498.

p. 294: *"In short, it's like a guerrilla war"*: Some scientists, notably Luigi Ferrucci of the NIA and The Blast, think that the job of cellular maintenance and repair is also under the domain of the immune system. So a virus like CMV would obviously help accelerate our aging, while the ability to restore a functioning thymus gland could have vast effects on the aging process generally. (People who take HGH also report that their thymus glands regrow, but they do not seem to work very well.)

p. 294, *"it was born in an epiphany that de Grey experienced"*: The epiphany is described at the beginning of de Grey's book, *Ending Aging*.

p. 296, *"a huge, $9 million EU-funded research initiative called ThymiStem":* The project has a website, http://www.thymistem.org; the 2014 paper can be found at N. Bredenkamp, C. S. Nowell, and C. C. Blackburn, "Regeneration of the Aged Thymus by a Single Transcription Factor," *Development* 141, no. 8 (April 2014): 1627-37.

p. 299, *"food and energy prices would go 'through the roof,'":* Randall Kuhn's SENS6 talk can be found here in its entirety: www.youtube.com/watch?v=F2s-RdkAB_4.

p. 301, *"it is a mainspring of human activity":* Ernest Becker, *The Denial of Death* (New York: Free Press, 1973).

Appendix: Things That Might Work

Resveratrol:

Poulsen Morten Møller et al. "Resveratrol in metabolic health: an overview of the current evidence and perspectives," in *Annals of the New York Academy of Sciences.* 1290 (2013) 74–82.

Hector et al. "The effect of resveratrol on longevity across species: a meta-analysis." *Biology Letters* (2012) 8, 790–793. Published online June 20, 2012. Basically refutes the whole notion that resveratrol increases lifespan in any animal.

Mattison, J. A., M. Wang, M. Bernier, J. Zhang, S. S. Park, S. Maudsley, S. S. An, et al. "Resveratrol Prevents High Fat/Sucrose Diet-Induced Central Arterial Wall Inflammation and Stiffening in Nonhuman Primates." *Cell Metabolism* 20, no. 1 (July 1 2014): 183-90.

Walle, T. "Bioavailability of Resveratrol." *Annals of the New York Academy of Sciences* 1215 (January 2011): 9-15.

Rossi D., A. Guerrini, R. Bruni, E. Brognara, M. Borgatti, R. Gambari, S. Maietti, G. Sacchetti. "trans-Resveratrol in Nutraceuticals: Issues in Retail Quality and Effectiveness." *Molecules.* 2012; 17(10):12393-12405. Addresses the lack of resveratrol in many alleged resveratrol supplements.

Weintraub, Arlene. "Resveratrol: The Hard Sell on Anti-Aging." *Businessweek,* July 29, 2009. On the booming resveratrol supplement industry.

Alcohol/Red Wine:

Much of the relevant red-wine research is summed up elegantly in J. P. Broustet, "Red Wine and Health," *Heart* 1999;81:459-460 (previously cited). Importantly, these were probably small, European glasses of wine, four ounces rather than the six ounce American pour.

These guys reviewed ten large studies and found that ALL alcoholic drinks conferred some level of protection against cardiovascular disease. Rimm, Eric B.,

Arthur Klatsky, Diederick Grobbee, and Meir J Stampfer. "Review of Moderate Alcohol Consumption and Reduced Risk of Coronary Heart Disease: Is the Effect Due to Beer, Wine, or Spirits?" *British Medical Journal* Vol. 312,731. 1996.

This study is of particular note, because it was done in Bordeaux and because it found a tremendous protective effect of red wine against Alzheimer's disease. The authors conclude, "There is no medical rationale to advise people over 65 to quit drinking wine moderately, as this habit carries no specific risk and may even be of some benefit for their health." (Also, they define "moderate" drinking as 3-4 glasses/day.) Orgogozo, J. M., J. F. Dartigues, S. Lafont, L. Letenneur, D. Commenges, R. Salamon, S. Renaud, and M. B. Breteler. "Wine Consumption and Dementia in the Elderly: A Prospective Community Study in the Bordeaux Area." [In English.] *Revue Neurologique (Paris)* 153, no. 3 (April 1997): 185-92.

More recent research is also suggesting that eating fat is not necessarily as bad for you as was thought when the "French Paradox" was coined in 1987—so the Paradox might not be such a paradox after all. Much of this is summed up in Teicholz, Nina. *The Big Fat Surprise: Why Butter, Meat, and Cheese Belong in a Healthy Diet.* 1st Simon & Schuster hardcover ed. New York: Simon & Schuster, 2014.

Coffee:

The big review: Freedman, N. D., Y. Park, C. C. Abnet, A. R. Hollenbeck, and R. Sinha. "Association of Coffee Drinking with Total and Cause-Specific Mortality." *New England Journal of Medicine* 366, no. 20 (May 17 2012): 1891-904.

Another big European study here: Floegel, Anna, Tobias Pischon, Manuela M. Bergmann, Birgit Teucher, Rudolf Kaaks, and Heiner Boeing. "Coffee Consumption and Risk of Chronic Disease in the European Prospective Investigation into Cancer and Nutrition (Epic)–Germany Study." *The American Journal of Clinical Nutrition* 95, no. 4 (April 1, 2012): 901-08.

Thoughtful editorial to accompany the above: Lopez-Garcia, Esther. "Coffee Consumption and Risk of Chronic Diseases: Changing Our Views." *The American Journal of Clinical Nutrition* 95, no. 4 (April 1, 2012): 787-88.

Curcumin:

See notes for p.85, which are on p. 328.

"Life Extension Mix":

Spindler, S. R., P. L. Mote, and J. M. Flegal. "Lifespan Effects of Simple and Complex Nutraceutical Combinations Fed Isocalorically to Mice." *Age* (Dordr) 36, no. 2 (April 2014): 705-18.

Metformin:

Martin-Montalvo, A., E. M. Mercken, S. J. Mitchell, H. H. Palacios, P. L. Mote, M. Scheibye-Knudsen, A. P. Gomes, et al. "Metformin Improves Healthspan and Lifespan in Mice." *Nature Communications* 4 (2013): 2192.

DeCensi, Andrea, Matteo Puntoni, Pamela Goodwin, Massimiliano Cazzaniga, Alessandra Gennari, Bernardo Bonanni, and Sara Gandini. "Metformin and Cancer Risk in Diabetic Patients: A Systematic Review and Meta-Analysis." *Cancer Prevention Research* 3, no. 11 (November 1, 2010): 1451-61.

Kasznicki, J., A. Sliwinska, and J. Drzewoski. "Metformin in Cancer Prevention and Therapy." *Annals of Translational Medicine* 2, no. 6 (June 2014): 57.

Vitamin D

Brunner, R. L., B. Cochrane, R. D. Jackson, J. Larson, C. Lewis, M. Limacher, M. Rosal, S. Shumaker, and R. Wallace. "Calcium, Vitamin D Supplementation, and Physical Function in the Women's Health Initiative." *Journal of the American Diet Association* 108, no. 9 (September 2008): 1472-9.

Bjelakovic, G., L. L. Gluud, D. Nikolova, K. Whitfield, J. Wetterslev, R. G. Simonetti, M. Bjelakovic, and C. Gluud. "Vitamin D Supplementation for Prevention of Mortality in Adults." *Cochrane Database Systematic Reviews* 1 (2014): CD007470.

Holick, M. F. *The Vitamin D Solution: A 3-Step Strategy to Cure Our Most Common Health Problem,* (New York: Hudson Street Press, 2010).

Gordon Lithgow talks about the vitamin D story here: http://vimeo.com/channels/thebuck/67168737.

Aspirin and Ibuprofen

There are lots and lots of studies on aspirin and anti-inflammatories. The Womens' Health Initiative study (same one that shot down hormone replacement) found a strong association between aspirin use and reduced mortality: Berger, J. S., D. L. Brown, G. L. Burke, A. Oberman, J. B. Kostis, R. D. Langer, N. D. Wong, and S. Wassertheil-Smoller. "Aspirin Use, Dose, and Clinical Outcomes in Postmenopausal Women with Stable Cardiovascular Disease: The Women's Health Initiative Observational Study." *Circulatory and Cardiovascular Quality and Outcomes* 2, no. 2 (March 2009): 78-87.

Strong, Randy, Richard A. Miller, et al. "Nordihydroguaiaretic Acid and Aspirin Increase Lifespan of Genetically Heterogeneous Male Mice." *Aging Cell* 7, no. 5 (2008): 641-50. Found that it made mice live longer—males only.

Vlad, S. C., D. R. Miller, N. W. Kowall, and D. T. Felson. "Protective Effects of NSAIDs on the Development of Alzheimer Disease." *Neurology* 70, no. 19 (May 6, 2008): 1672-7.

Kale

Only if you like the taste.

Further Reading

Agus, David. *A Short Guide to a Long Life.* New York: Simon & Schuster, 2014.

Agus, David, and Kristin Loberg. *The End of Illness.* 1st Free Press hardcover ed. New York: Free Press, 2012.

Alexander, Brian. *Rapture: How Biotech Became the New Religion.* 1st ed. New York: Basic Books, 2003.

Aminoff, Michael J. *Brown-Séquard: An Improbable Genius Who Transformed Medicine.* New York: Oxford University Press, 2010.

Arrison, Sonia. *100 Plus: How the Coming Age of Longevity Will Change Everything: From Careers and Relationships to Family and Faith.* New York: Basic Books, 2011.

Austad, Steven N. *Why We Age: What Science Is Discovering About the Body's Journey through Life.* New York: J. Wiley & Sons, 1997.

Becker, Ernest. *The Denial of Death.* New York: Free Press, 1973.

Boyle, T. Coraghessan. *The Road to Wellville.* New York: Viking, 1993.

Brock, Pope. *Charlatan: America's Most Dangerous Huckster, the Man Who Pursued Him, and the Age of Flimflam.* 1st ed. New York: Crown Publishers, 2008.

Buettner, Dan. *The Blue Zones: 9 Lessons for Living Longer from the People Who've Lived the Longest.* 2nd ed. Washington, D.C.: National Geographic, 2012.

Campbell, T. Colin, and Thomas M. Campbell. *The China Study: The Most Comprehensive Study of Nutrition Ever Conducted and the Startling Implications for Diet, Weight Loss and Long-Term Health.* Dallas: BenBella Books, 2005.

Comfort, Alex. *The Process of Ageing.* Signet Science Library. New York: New American Library, 1964.

Comfort, Alex. *Ageing, the Biology of Senescence.* Rev. and reset ed. New York: Holt, 1964.

Cornaro, Luigi, Joseph Addison, Francis Bacon, and William Temple. *The Art of Living Long; a New and Improved English Version of the Treatise.* Milwaukee: W. F. Butler, 1905.

Cowdry, Edmund Vincent, and Edgar Allen. *Problems of Ageing; Biological and Medical Aspects.* 2nd ed. Baltimore: The Williams & Wilkins Company, 1942.

Critser, Greg. *Eternity Soup: Inside the Quest to End Aging.* 1st ed. New York: Harmony Books, 2010.

Crowley, Chris, and Henry S. Lodge. *Younger Next Year: A Guide to Living Like 50 until You're 80 and Beyond.* New York: Workman Publishing, 2004.

De Grey, Aubrey D. N. J., and Michael Rae. *Ending Aging: The Rejuvenation Breakthroughs That Could Reverse Human Aging in Our Lifetime.* 1st ed. New York: St. Martin's Press, 2007.

Finch, Caleb. *Longevity, Senescence, and the Genome.* The John D. and Catherine T. MacArthur Foundation Series on Mental Health and Development. Chicago: University of Chicago Press, 1990.

Finch, Caleb, and Leonard Hayflick. *Handbook of the Biology of Aging.* The Handbooks of Aging. New York: Van Nostrand Reinhold Co., 1977.

Gawande, Atul. *Being Mortal: Medicine and What Matters in the End.* 1st ed. New York: Metropolitan Books: Henry Holt & Company, 2014.

Gruman, Gerald J. *A History of Ideas About the Prolongation of Life.* Classics in Longevity and Aging Series. New York: Springer Publishing Co., 2003.

Guarente, Leonard. *Ageless Quest: One Scientist's Search for Genes That Prolong Youth.* Cold Spring Harbor, NY: Cold Spring Harbor Laboratory Press, 2003.

Hadler, Nortin M. *The Last Well Person: How to Stay Well Despite the Health-Care System.* Montreal; Ithaca: McGill-Queen's University Press, 2004.

Haldane, J. B. S. *New Paths in Genetics.* London: G. Allen & Unwin Ltd., 1941.

Hall, Stephen S. *Merchants of Immortality: Chasing the Dream of Human Life Extension.* Boston: Houghton Mifflin, 2003.

Hayflick, Leonard. *How and Why We Age.* 1st ed. New York: Ballantine Books, 1994.

Holick, M. F. *The Vitamin D Solution: A 3-Step Strategy to Cure Our Most Common Health Problem.* New York: Hudson Street Press, 2010.

Jacobs, A. J. *Drop Dead Healthy: One Man's Humble Quest for Bodily Perfection.* 1st Simon & Schuster hardcover ed. New York: Simon & Schuster, 2012.

Jerome, John. *Staying With It.* New York: Viking Press, 1984.

Kerasote, Ted. *Pukka's Promise: The Quest for Longer-Lived Dogs.* Boston: Houghton Mifflin Harcourt, 2013.

Kurzweil, Ray, and Terry Grossman. *Transcend: Nine Steps to Living Well Forever.* Emmaus, PA: Rodale, 2009.

Lakatta, E. G. "Arterial and Cardiac Aging: Major Shareholders in Cardiovascular Disease Enterprises: Part III: Cellular and Molecular Clues to Heart and Arterial Aging." *Circulation* 107, no. 3 (Jan 28 2003): 490-7.

Lakatta, E. G., and D. Levy. "Arterial and Cardiac Aging: Major Shareholders in Cardiovascular Disease Enterprises: Part II: The Aging Heart in Health: Links to Heart Disease." *Circulation* 107, no. 2 (Jan 21 2003): 346-54.

———. "Arterial and Cardiac Aging: Major Shareholders in Cardiovascular Disease Enterprises: Part I: Aging Arteries: A 'Set up' for Vascular Disease." *Circulation* 107, no. 1 (Jan 7 2003): 139-46.

Lieberman, Daniel. *The Story of the Human Body: Evolution, Health, and Disease.* 1st ed. New York: Pantheon Books, 2013.

Life, Jeffry S. *The Life Plan: How Any Man Can Achieve Lasting Health, Great Sex, and a Stronger, Leaner Body.* 1st Atria Books hardcover ed. New York: Atria Books, 2011.

Masoro, Edward J., and Steven N. Austad. *Handbook of the Biology of Aging.* The Handbooks of Aging. 6th ed. Amsterdam; Boston: Elsevier Academic Press, 2006.

McCay, Jeanette B. *Clive McCay, Nutrition Pioneer: Biographical Memoirs by His Wife.* Charlotte Harbor, FL: Tabby House, 1994.

Mitchell, Stephen. *Gilgamesh: A New English Version.* New York: Free Press, 2004.

Moalem, Sharon, and Jonathan Prince. *Survival of the Sickest: A Medical Maverick Discovers Why We Need Disease.* 1st ed. New York: William Morrow, 2007.

Mosley, Michael, and Mimi Spencer. *The Fast Diet: Lose Weight, Stay Healthy, and Live Longer with the Simple Secret of Intermittent Fasting.* 1st Atria Books hardcover ed. New York: Atria Books, 2013.

Mukherjee, Siddhartha. *The Emperor of All Maladies: A Biography of Cancer.* 1st Scribner hardcover ed. New York: Scribner, 2010.

Nuland, Sherwin B. *The Art of Aging: A Doctor's Prescription for Well-Being.* 1st ed. New York: Random House, 2007.

Oeppen, J., and J. W. Vaupel. "Demography. Broken Limits to Life Expectancy." *Science* 296, no. 5570 (May 10, 2002): 1029-31.

Olshansky, Stuart Jay, and Bruce A. Carnes. *The Quest for Immortality: Science at the Frontiers of Aging.* New York: Norton, 2001.

Poynter, Jane. *The Human Experiment: Two Years and Twenty Minutes Inside Biosphere 2.* New York, Berkeley, CA: Thunder's Mouth Press; Distributed by Publishers Group West, 2006.

Ridley, Matt. *Genome: The Autobiography of a Species in 23 Chapters.* 1st U.S. ed. New York: HarperCollins, 1999.

Shock, Nathan W., and Gerontology Research Center (U.S.). *Normal Human Aging: The Baltimore Longitudinal Study of Aging.* NIH Publication. Baltimore, Md.; Washington, D.C.: U.S. Dept. of Health and Human Services, Public Health Service, National Institutes of Health, National Institute on Aging. For sale by the Supt. of Docs., U.S. G.P.O., 1984.

Shteyngart, Gary. *Super Sad True Love Story: A Novel.* 1st ed. New York: Random House, 2010.

Snowdon, David. *Aging with Grace: What the Nun Study Teaches Us About Leading Longer, Healthier, and More Meaningful Lives.* New York: Bantam Books, 2001.

Stipp, David. *The Youth Pill: Scientists at the Brink of an Anti-Aging Revolution.* New York: Current, 2010.

Taubes, Gary. *Good Calories, Bad Calories: Challenging the Conventional Wisdom on Diet, Weight Control, and Disease.* 1st ed. New York: Knopf, 2007.

———. *Why We Get Fat and What to Do About It.* 1st ed. New York: Alfred A. Knopf, 2011.

Teicholz, Nina. *The Big Fat Surprise: Why Butter, Meat, and Cheese Belong in a Healthy Diet.* 1st Simon & Schuster hardcover ed. New York: Simon & Schuster, 2014.

Varady, Krista. *The Every-Other-Day Diet: The Diet That Lets You Eat All You Want (Half the Time) and Keep the Weight Off.* 1st ed. New York: Hyperion, 2013.

Weinberger, Eliot. *Karmic Traces, 1993-1999.* New Directions Paperbook. New York: New Directions Books, 2000.

Weiner, Jonathan. *Long for This World: The Strange Science of Immortality.* 1st ed. New York: Ecco, 2010.

Weintraub, Arlene. *Selling the Fountain of Youth: How the Anti-Aging Industry Made a Disease out of Getting Old, and Made Billions.* New York: Basic Books, 2010.

Whitehouse, Peter J., and Daniel George. *The Myth of Alzheimer's: What You Aren't Being Told About Today's Most Dreaded Diagnosis.* 1st ed. New York: St. Martin's Press, 2008.

Index

B

C

D

E

F

I

J

K

L

N

O

P

R

S

Y

Z

About The Author

Glenn Acker

Bill Gifford is a contributing editor of *Outside* magazine and has written extensively on science, sports, health and fitness for *Wired*, *Men's Journal*, *Slate*, *New Republic*, *Men's Health*, and *Bicycling*, among other publications. He has been an editor at *Men's Journal* and *Philadelphia* magazines, and his work has been anthologized in *Best American Sports Writing*. Gifford is also the author of *Ledyard: In Search of the First American Explorer*. He lives in Mount Gretna, Pennsylvania, and the person he most admires is his grandmother, Doris, who is nearly 100 years old and eats a Danish pastry every morning. He intends to get there, too. Or die trying.